RFID and Beyond

RFID and Beyond

GROWING YOUR BUSINESS THROUGH REAL WORLD AWARENESS

Claus Heinrich

WILEY

Wiley Publishing, Inc.

Acquisitions Editor: Katie Mohr
Senior Development Editor: Kevin Kent
Development and Copy Editor: Rebecca Whitney
Editorial Manager: Mary Beth Wakefield
Vice President and Executive Group Publisher: Richard Swadley
Vice President and Publisher: Joe Wikert
Project Coordinator: Erin Smith
Graphics and Production Specialists: Carrie A. Foster, Denny Hager, Jennifer Heleine
Quality Control Technicians: Amanda Briggs, Leeann Harney, Jessica Kramer
Proofreading and Indexing: TECHBOOKS Production Services

RFID and Beyond: Growing Your Business Through Real World Awareness
Published by
Wiley Publishing, Inc.
10475 Crosspoint Boulevard
Indianapolis, IN 46256
www.wiley.com

Copyright © 2005 by Wiley Publishing, Inc., Indianapolis, Indiana
Published simultaneously in Canada

Library of Congress Cataloging-in-Publication Data

Heinrich, Claus E.
  RFID and beyond : growing your business through real world awareness / Claus Heinrich.
     p. cm.
  Includes index.
  ISBN 0-7645-8335-2 (pbk.)
  1. Inventory control--Automation. 2. Radio frequency identification
systems. I. Title.
  TS160.H45 2005
  658.7'87--dc

                        222004030558

Manufactured in the United States of America
10 9 8 7 6 5 4 3 2 1
1CW/ST/QS/QV/IN

Contents

PART II: EXPERTS IN REAL WORLD
            AWARENESS   191

Preface

RFID, RFID, RFID. Seemingly everywhere you turn in industries as diverse as cattle ranching, retailing, pharmaceuticals, medical devices, consumer products, and defense department logistics, Radio Frequency IDentification (RFID) is a hot topic. Reading the trade press today, you might get the impression that everything will be tagged. Large companies, like Wal-Mart, METRO Group, and the U.S. Department of Defense, are the masters of distribution channels and are insisting that RFID be used. Brand masters, like Procter & Gamble and Colgate, are readying their factories and supply chains to use RFID in ways that allow them to become more responsive to consumer demand and to increase efficiency and flexibility to new levels to meet the demands of an increasingly competitive market. Technology vendors of all sorts, from software vendors to systems integrators to chip makers, are ready to hit the ground running with major implementation projects.

Today, you can hardly find an executive who does not have an opinion on RFID. The hype has reached such a fever pitch that Gartner has released a paper recommending that IT executives prepare for disillusionment. RFID projects, Gartner says, cannot meet expectations.

On the other hand, there are few other topics where there is such an obvious need for basic information.

Wise IT executives avoid disappointment by not just following the trend, but instead leveraging this new technology where it creates real business value.

But even the harshest critics of RFID, those who point out all the difficulties with getting RFID readers to work properly, with changing the way that software works and business processes are designed, don't deny that RFID and similar technologies are perhaps the biggest thing to hit the IT world since the Internet, and may be as profound in their ramifications. Some consumer and advocacy groups fear the potential negative social effects so much that they want to stop RFID in its tracks.

What all this hype, activity, and discourse misses is that RFID is not really the point. RFID is just one instance of a bigger concept. RFID will be improved, adjusted, and perhaps one day replaced. But the way that RFID create value—its ability to sense information about the real world in a wholesale manner, send that information to intelligent systems through network connectivity of all sorts, and then respond in various ways that create value—is something new under the sun. Our name for this larger concept is Real World Awareness. This book is a first stab at creating a general overview of the nature and implications of that idea.

What business executives and technologists in all industries need is a framework, a taxonomy, a mental model, and an outline of how to understand the changes that Real World Awareness will bring to their businesses and how to organize their learning. This book provides such a framework and explains how to think about Real World Awareness, how the technology works, and how it will change the way business works. The book includes dispatches from the field about early implementation projects and discusses social and political issues that are being debated. Interviews from leading executives are included throughout the book, as well as contributions from leading legal and academic thinkers.

WHO SHOULD READ THIS BOOK?

The goal of this book is to explain to more or less anybody involved in business what the fuss concerning RFID is about. Rather than explain just one technology, we expand the idea of RFID into the more general concept of Real World Awareness. Senior executives should find enough information to enable them to converse with their most technical-minded employees. Technologists will find out why businesspeople care so much about the functionality of Real World Awareness technologies.

Information technology executives of all stripes will find explanations that enable them to better communicate. In general, this book is for people who want not only to know about Real World Awareness and RFID, but also to do something about them.

CHAPTER SUMMARIES

RFID and Beyond has 10 chapters, presented in two parts. Part I, "Real World Awareness in Business," comprises Chapters 1 through 6, which tell the story of how Real World Awareness will change the world of business.

- **Chapter 1, "Business Navigation and Real World Awareness,"** traces the history of Real World Awareness in aviation and shows you how the same stages of evolution are now taking place in the business world.
- **Chapter 2, "How Real World Awareness Will Change Your Business,"** shows you how Real World Awareness enables new business paradigms, like the adaptive business network, to work in an optimal manner. The consumer-driven supply network is examined in detail, and the ways that business processes will likely change are surveyed.
- **Chapter 3, "The Technologies of Real World Awareness,"** examines dimensions of the different components, such as RFID tags, smart cards, and sensors, and shows you how this technology works by using examples such as the Metro Group Future Store.
- **Chapter 4, "Business Process Design and Optimization,"** explains how processes and business models will change as Real World Awareness provides more information. Different paradigms, such as predictive maintenance, are described, as well as the analytical techniques that will be used as automation expands.
- **Chapter 5, "Implementing Real World Awareness,"** surveys the challenges that early adopters have been facing during implementation and the methods they have used to overcome them.
- **Chapter 6, "People, Privacy, Politics,"** covers the social and political issues that are being discussed as techniques such as RFID become more widely noticed.

Part II, "Experts in Real World Awareness," consists of Chapters 7 through 10, which are contributions from scholars and analysts who address key issues related to Real World Awareness.

- **Chapter 7, "Mastering the Legal Challenges,"** surveys the legal issues that must be addressed as implementations of Real World Awareness systems branch out.
- **Chapter 8, "The Impact of RFID on Supply Chain Efficiency,"** looks at the future of supply chain management and adaptive business networks.
- **Chapter 9, "Exploding Edges and Potential for Disruption,"** examines how the explosion of devices at the edge of the network could have revolutionary effects.
- **Chapter 10, "The X Internet Unleashes Real World Awareness Services Revolution,"** examines the way that services offerings will be enhanced by Real World Awareness.

ACKNOWLEDGMENTS

My name, Claus Heinrich, is on the cover of this book because it reflects my thinking and vision for Real World Awareness. But the creation of this book benefited greatly from a team made up of people I have worked with for years and some I just met while working on this project.

Robert Cummings, Stefan Schaffer, Ralph Schneider, and Alexander Zeier helped immensely with detailed reviews and suggestions on several chapters. Alexander also acted as project manager and kept the book moving quickly from beginning to end. Dan Woods and his team from the Evolved Media Network provided valuable assistance in editing and also helped create the information graphics that make this book a lively read. I would also like to thank Antonia Ashton for her guidance and counsel about shaping this book to communicate with the broadest possible audience, and for her able stewardship of marketing and promotional activities. As always, I am deeply grateful for the organizational skills and insight of Susanne Schaehfer, who performed brilliantly as reviewer, coordinator, and problem solver for this demanding endeavor.

But my most profound thanks go to the executives, professors, and analysts who contributed to this book. They became my partners as authors for many different sections, and it is their contributions that take this book beyond the realm of the average business book by informing it with experience garnered from decades spent on the technology battlefield. Readers will likely find the contributions from the following executives to be some of the best parts of this book:

- Peter Bauer, Executive VP, Chief Sales and Marketing Officer, and a Member of the Management Board, Infineon
- Michael Dell, Chairman and Founder, Dell Inc.
- Ron Dennis, Chairman and CEO, McLaren Group
- Charles Fine, Professor, Massachusetts Institute of Technology
- Pat Gelsinger, Senior Vice President, Chief Technology Officer, Intel Corporation
- Keith Harrison, Global Product Supply Officer, Procter & Gamble
- Stefan Lauer, Executive Board Member, Deutsche Lufthansa
- Jonathan Loretto, Capgemini's Executive Representative at EPCglobal and Global Technology Lead on RFID for Capgemini Group
- Zygmunt Mierdorf, Chief Information Officer, METRO Group
- Pekka Ala-Pietilä, President, Nokia Corporation
- Navi Radjou, Vice President, Enterprise Applications Team, Forrester Research
- Viola Schmid, Professor, Darmstadt University of Technology
- David Simchi-Levi, Professor, Massachusetts Institute of Technology
- Ed Toben, Chief Information Officer, Colgate
- Ray Valeika, former Senior Vice President for Technical Operations, Delta Air Lines (retired October 1, 2004)
- Juergen Weber, Chairman of the Board, Lufthansa
- Klaus Zumwinkel, CEO and Chairman of the Board of Management, Deutsche Post World Net

PART I

REAL WORLD
AWARENESS
IN BUSINESS

1

Business Navigation and Real World Awareness

odern aviation is the complete realization of Real World
Awareness. As a passenger, when you board a flight from
Frankfurt to New York, your thoughts may roam to which
movie will be played. But, in the cockpit, the pilot and copilot are in an
intense conversation with a computer that is making sure that all the
sensors that keep track of the plane's engines, speed, altitude, radar, GPS,
and fuel levels are in working order. The pilots are also checking higher-
order functions that use these sensors in intelligent ways. The autopilot,
the ground-proximity warning system, the landing system, the communications system, and the navigation systems are all run through diagnostic routines. As the plane takes off, sensors on the engines are taking
dozens of readings that will later be sent via satellite to a central computer that is monitoring the operation of the engine and looking for
patterns that indicate problems or inefficiencies. While you are wonder-
ing whether your favorite Chardonnay will be available with dinner, the
pilot is managing and monitoring all these automated systems and waiting for them to indicate any problems so that the right response can be
executed.

This modern flight is the essence of the concept of Real World
Awareness, which is the topic of this book. Data from sensors provide
information about the real world. Automated systems use that data to

carry out complicated tasks. Advanced data analysis looks for patterns in the future. The pilot is in the center of it all, monitoring, managing, and handling exceptional cases. All this is at the service of the customer, who, although unaware of the means, knows that the flight is safe, well managed, and available for an attractive price. This finely tuned machine did not start out this way, but was the result of decades of evolution.

In the early days of aviation, well into the 1930s, flying a plane was an art that required skill and intuition, not to mention a serious dose of bravery. The cockpit was sparse, with a joystick, a couple of levers here and there, and a good view. Speedometers and altimeters were all in the future then, and pilots flew planes by feeling based on experience. This method was called *flying by the seat of your pants* because the pilot's "rear end" was literally one of the most important sensors! Pilots wore thin clothing and sat on seats with little padding in order to better sense the vibrations of the plane. Experienced pilots would know what sort of vibrations were okay and which bumps and jiggles indicated a problem. Figure 1.1 shows the cockpit that Charles Lindbergh faced when he crossed the Atlantic in 1927.

Figure 1.1. Cockpit from the *Spirit of St. Louis*
©2002 Orbital Air, Inc. All rights reserved. www.OrbitalAir.com

As aviation technology progressed, more instruments were added to the cockpit to provide the sort of information that achieves Real World Awareness. At first, air-speed indicators, altimeters, and compasses appeared, followed eventually by radios, radar, and primitive flight-warning computers that sensed dangerous conditions. Figure 1.2 shows the cockpit of a DC-3, which added a significant number of instruments. The pilot was joined by a copilot, navigator, radio officer, and flight engineer, who was responsible for maintaining the machinery while flying. As technology advanced, computers changed cockpits dramatically, sometimes in ways initially resisted by pilots. Advanced navigation systems, fly-by-wire systems, and autopilots were all accepted after a struggle, which included fights over the elimination of the radio officer, navigator, and then the flight engineer, which gave aviation the modern two-person cockpit, also known as the "glass cockpit."

Figure 1.2. Cockpit from a DC-3

In today's planes, you find advanced navigation systems that have maps of the geography of the entire planet. Figure 1.3 shows you the glass cockpit of the Airbus 300 series, which was the first in a long line of computerized designs that skeptical pilots derisively called "the electronic pig." Autopilots have evolved to automate most of the tasks of flying a plane, including landing. Pilots no longer hold the plane on course with their hands, but, rather, manage the information presented to them by the computers, which warn pilots of upcoming threats and prevent them from executing maneuvers that would lead beyond the limits of the safe flight envelope. In the most advanced high-performance aircraft that fly at incredible speeds and have a minute tolerance for error, humans cannot even fly the plane. High-performance computers are required to perform most of the tasks of adjusting flight controls, such as flaps, at speeds of Mach 2 or higher.

Figure 1.3. Glass cockpit from an Airbus A300

A transformation in business is now under way that runs parallel to the evolution of aviation technology. Despite the application of advanced technology and business systems, too many business executives still fly their organizations by the seats of their pants. It is not because the business systems do not work, but rather because, until quite recently, even the most modern systems were still based on *assumptions,* not on real-time data about where goods are located, what state they are in, and other data that could provide an accurate map of reality. The information in these systems is only as accurate as the last inventory check or data entry into the system. Real-time essentially meant "as soon as someone let the computer know!" Without *real* real-time data, businesses face uncertainty. Business-oriented Real World Awareness techniques, such as Radio Frequency IDentification, are dramatically reducing the cost of automatically and instantly acquiring accurate information about almost every aspect of a business. The gap between the state of the virtual world in business systems and that of the real world, which is being managed, can be reduced dramatically. When this happens, immense value can be created by increasing efficiency, expanding automation, or opening the door to new lines of business. Like a pilot in a modern plane, with Real World Awareness, executives have a new job. Automated systems can sense the exact state of the real world and respond appropriately. When exceptional conditions arise, both pilots and executives have a wealth of information to help determine a response.

We can still hear the echo of the starting gun in the race to apply Real World Awareness. There is much more to learn than is now known about how to apply this powerful technique. What this book aims to do is to take a snapshot of where we are now and explain a coherent way of thinking about the important issues because the stakes in this race are incredibly high. Companies must figure it out for themselves. The winners will be those who learn how to adapt their way of doing business to the new possibilities that Real World Awareness brings.

REAL WORLD AWARENESS IN AVIATION

Flying by the seat of the pants is really not that bad of an idea, if you have no other choice. Pilots took off and landed, navigated by compasses or points on land, and got from here to there just fine. Good pilots could tell what was going on as their planes rumbled along, and many impressive feats of aviation were achieved.

But, although flying by the seat of the pants was an admirable skill, it was far from a well-defined, scientific process and did not lend itself to logical analysis, incremental and standardized improvement, or the development of safety guidelines and best practices. It was a matter of people doing the best they could with the available technology and experience. When more modern technology progressively arrived, aviation changed drastically.

The message for a Real World Aware business:

> Flying by the seat of your pants is acceptable only when you have no other option.

INSTRUMENTS AND AUTOMATION

The arrival of instruments to determine a plane's speed and altitude, its primitive weather measurements, and the performance of the engine changed flying from an intuitive skill to a process that was far more accurately defined and predictable. The speed at which a plane should be traveling at take-off and landing could be clearly specified. The proper readings of oil pressure, coolant temperature, and other parameters and indicators of stress on the engine could be easily documented. It didn't take long for the work of flying a plane to be expanded from one brave individual to a team designed to minimize risk.

As larger planes developed with multiple engines, a copilot joined the pilot to provide safety through redundancy. A flight engineer joined the crew to closely monitor the increasing amount of machinery and electronics on the plane. A navigator was also introduced to perform all the calculations that were required, because more primitive forms of navigation no longer worked for long flights during the day or night or at high altitudes. Radio officers were introduced to operate communications equipment on-board since early long-range communications were transmitted through Morse Code.

The number of instruments on the plane quickly ballooned and left the flight engineer with hundreds of points of data to keep track of as the plane flew along. The flight engineer's job and the crew's task of monitoring were made easier by an automatic warning system, which was an early example of today's automated computer analysis system, the flight-warning computer. The flight-warning computer was pro- grammed to monitor many key indicators on the plane and to emit a warning alert whenever certain thresholds were exceeded. This system provided assistance to the flight engineer and crew and served as a fail- safe device in case certain problems were missed. As the number of instruments on planes grew larger, the job of the flight engineer and crew became impossible without automation that analyzed values from hundreds of sensors and provided guidance to the flight engineer and crew about how to conduct analysis during the flight to ensure safety.

The message for a Real World Aware business:

> As the number of sources and the quantity of information grow, analysis must be automated.

PROCESS CHANGES IN THE COCKPIT

The flight-warning computer was just the beginning. Planes became much more complex. Jet engines, pressurized cabins, high-altitude flights, and radar all came with new technology that at first required care and special handling by the flight engineer and navigator, but eventually became highly integrated through automation.

Navigation was the first system to change dramatically. Even in the late 1970s, navigation still required the use of some ancient methods. On transcontinental flights, navigators would use a sextant to observe the stars through special slots in the ceiling of the cockpit. But then naviga- tion computers arrived that used a variety of methods, and eventually global positioning systems, to completely automate navigation and pro- vide a real-time view of the position of the plane and any deviation from the planned course. When these systems were introduced, reaction from pilots and navigators ranged from the suspicious to the downright hostile. On some airlines, even after the systems were introduced and in use, pilots and navigators refused to use them. Especially on flights that crossed the North Pole, pilots were afraid that the systems wouldn't work in the strange polar magnetic environment.

As it turned out, after a period of struggle, during which navigators continued to fly as part of the crew and most often did nothing, the automated navigation systems were finally accepted and the flight crew was reduced to three people: the pilot, copilot, and flight engineer. The navigator was eliminated. The radio officer was also displaced from the cockpit, as modernized telecommunications equipment simplified operations by pilots.

Automation quickly overtook the flight engineer as well, as new cockpits were introduced to allow the pilot and copilot to monitor and conduct operations with a variety of systems that automated the analysis of the readings coming from hundreds, and eventually thousands, of sensors all over the plane.

These developments hold two messages for a Real World Aware business:

- More information at first improves the current way of doing things and then creates completely new ones.
- Some changes may not be well received.

FLY BY WIRE

Aviation is inherently risky, which leads to engineering practices that emphasize redundancy. Components that may appear only once in a car may be replicated three or four times in an airplane. When one component fails, another takes over. If that one then fails, another is waiting, and so on. This system reduces the risk from the expected failure of components to the degree that pilots have time to land the plane before any catastrophic failure occurs.

As the number of sensors on a plane grew and the size of a plane grew, the number of cables connecting sensors to electronics grew even faster. This growth became a problem for reliability because many thousands of cables had to be connected properly, and for weight as the length and number of the cables added up to a significant load.

Airbus pioneered a solution, named *fly by wire*. This architecture replaced the thousands of electromechanical connections between the sensors and instruments and control mechanism in the cockpit with an electronic bus, or hub-and-spoke, design that could carry all the data from the multitude of sensors to the computers in the cockpit. This bus was similar to a superhighway for data that was used by all the sensors. This architecture replaced the complexity and heavy weight of thousands of wires with a much smaller and compact set of cables for the bus. Each sensor connected to the bus sent and received data from the cockpit.

One effect of this solution was an increase in standardization among sensors, which then all had to be constructed to use the bus architecture. Airbus took another leap forward in the standardization of cockpits in its A300 series of planes, which were built to a standardized design. Almost all planes in the A300 series, and many of its successors, use the same design for the instrumentation, display, and controls in the cockpit so that a pilot can learn to fly another plane in the series after only a couple of days of training. Some other aircraft manufacturers create new cockpit designs for each model, which means that pilots trained on one model may need as many as three months of training to fly another.

The message for a Real World Aware business from these developments:

- More sensors demand better connectivity.
- More sensors make standardization even more vital.

THE PILOT REDEFINED

Automation continued to surround the pilot. Ground-proximity warning systems would activate aural alarm signals if the plane was closing fast on terrain or a mountain or any obstacle so that the pilot could return to a safe altitude. The autopilot function, which maintains course and height according to a flight plan and automatically adjusts the direction and speed of an aircraft, became even more powerful. Key elements of a pilot's job, such as landing the plane, became highly automated so that the pilot could, under normal conditions, play the role of a highly intelligent failsafe device who would meticulously monitor everything and take over in case automated systems did not function properly or could not handle difficult conditions.

This increase in automation has profoundly changed the role of the pilot. No longer are the physical movements of the pilot directing the plane for most of the flight. The pilot has become an executive of the aircraft, a manager who defines the task that the automated systems should carry out and then manages any problems or exceptional conditions that arise. Pilots now have to understand the basic mechanics of the airplane, the control techniques with or without automation, and the operation of the automated systems controlling the plane so that they can be properly programmed. When problems occur, the pilot now resorts to the use of advanced diagnostic systems that provide insight and alternative corrective actions to handle what is happening on the plane. The entire suite of automation helps the pilot to visualize, by using graphics, the entire spectrum of operations of the aircraft.

The message of this transformation for a Real World Aware business:

Increased automation leads to management by exception and insight.

New Possibilities for Aviation

The discussion in this chapter has led to the two-person operation of a highly automated cockpit, as summarized in Figure 1.4. Instruments sense the state of the aircraft. Technology such as the flight-warning computer automates the analysis of the sensors. The fly-by-wire system provides connectivity between the sensors and the response mechanisms that automate complex processes such as navigation or, in the case of the autopilot, entirely change the way the plane is flown.

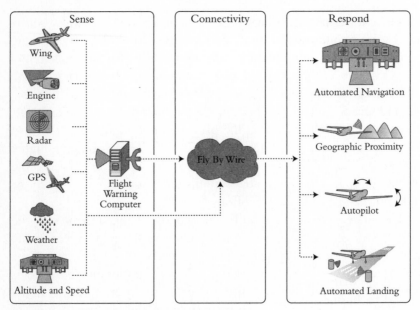

Figure 1.4. Real World Awareness in Aviation

Much of the action in aviation now is focused on reducing weight by replacing aluminum with much lighter carbon fiber in the construction of the fuselage and wings. Efforts are continuing to squeeze what may be the last 10 percent of efficiency from jet engines. The next wave of advances in automation is much more likely to affect the way planes interact with each other. The present tolerance for distance and height separation between planes is based on the capabilities of older-vintage planes. Scheduling algorithms, and the way that planes identify themselves and report on their conditions, can be improved. The amount of time spent in holding patterns over the Frankfurt airport each year, for example, accounts for enough fuel to make 200 trips to New York. This situation is a prime target for the next wave of automation.

The message for a Real World Aware business:

> Intelligent automated parts of a system can lead to dramatic new levels of efficiency, coordination, and collaboration in an adaptive business network.

Although aviation is a fascinating topic on its own, its history and complexity have served to present in this book, in an understandable way, the themes that companies will struggle with as Real World Awareness becomes an urgent concern. The next section examines how these themes will be addressed in the world of business.

APPLYING REAL WORLD AWARENESS TO BUSINESS

Real World Awareness applies to aviation, where the laws of physics rule in a fairly clear and straightforward manner. The world of business is more "fuzzy." The dynamics of markets, the complexities of organizational and human relationships, and the incessant pace of change make for a situation that is much less stable and a set of problems that is harder to solve. The fact is that in most businesses, there will always be some flying by the seat of the pants. The goal is to do as little of it as possible. Business executives need to make decisions based as much as possible on facts rather than on assumptions. The executives need to sense and respond to business conditions, make plans and execute them, and—most of all—learn from their experience.

Aviation is concerned with safety, and the goal of business is growth and long-term survival. In both cases, Real World Awareness is a requirement for optimal performance. In aviation, Real World Awareness allows pilots to fly safely at night, or into clouds, knowing whether a hailstorm or just water vapor lies ahead. Optimal routes can be calculated to save fuel or to avoid close calls with other planes. The route to be flown and the terrain of the entire world are continuously displayed to every pilot in the cockpit. This display presents all ground features, and the prevailing weather conditions, through digital maps in ground-proximity computers.

In business, Real World Awareness provides detailed information that transforms a vague seat-of-the-pants sense of what is going on in the huge variety of customer interactions from marketing, sales, and service—and the complex operations of a supply chain or manufacturing shop floor—into a precise real-time view of exactly what is happening in every important process. The availability of more information will transform business just as it has transformed aviation.

A similar pattern applies to the evolution of Real World Awareness in both aviation and business, as shown in Table 1.1. More information is collected, processes are improved and automated, and eventually new systems change the paradigm. Sensors provide information about the real world, which is then connected to automated mechanisms for analysis and response.

In aviation, these stages took decades as aviation engineers, aided by technological advances, improved the design and engineering for each successive model of aircraft. In business, the progress has been faster, less steady, and more lurching as companies have rushed to take advantage of mainframes, and then minicomputers, and then personal computers, and then client-server technology, and then applications ranging from spreadsheets to ERP systems. The Internet, of course, was the most dramatic recent driver of change.

For business, the arrival of affordable Real World Awareness technologies such as RFID, the wide availability of both wired and wireless networks, and the development of advanced systems for automation and analysis is now opening up a new world of opportunities.

Deutsche Lufthansa AG is one of the world's leading airlines. Alongside its core business of passenger and freight transport, Lufthansa offers its customers a number of specialized services related to air travel. More than 90,000 employees from 150 countries have made Lufthansa not only one of the world's leading aviation companies, but also an employer with a particularly international focus. Lufthansa is among Germany's 20 largest and most popular employers.

Juergen Weber

Chairman of the Board

Deutsche Lufthansa AG

Born in 1941, Juergen Weber joined Lufthansa in 1967 and moved through many different engineering and management positions on his way to the chairmanship of Lufthansa's executive board in 1991. During his long career, Juergen was instrumental in driving technical innovations in aircraft maintenance, for which he has received many awards. He also led Lufthansa through privatization and the creation of the Star Alliance. In June 2003, Juergen was elected chairman of Lufthansa's supervisory board.

Q&A with Juergen Weber, Chairman of the Board, Lufthansa AG:

Q: Aviation has pioneered the implementation of Real World Awareness. What is the next frontier for improvement in aviation?

A: We're approaching the end of the optimization curve of the "hardware" of flying, such as engines. It will be costly to push their efficiency any further. But a number of other great things can happen. One of the most promising areas is air traffic control and flight systems. The current procedures partially date back to the propeller era when navigation was done manually, with a precision of plus or minus 1 kilometer. Now, precision is very high, and computers can communicate with each other, which is not sufficiently exploited. Systems do not interact efficiently, and the result you get is holding circles, delays, and unnecessary fuel burn. Integrated management of processes will make sure that when you depart from Frankfurt your landing slot in New York JFK is already guaranteed, with strong winds perhaps as the only variable. This will reduce costs and improve safety and punctuality.

Q: You have spent much of your professional career on helping to set standards. Why are they so important?

A: Standards ensure the smooth running and safety of the airline industry. International Air Transport Association has taken a lot of care to standardize key components. If an airplane uses an automatic landing system in Frankfurt, it must be able to land with the same system in New York, in Singapore, and in Sydney as well.

When airplane cockpits differ in their layout, in the way switches and levers are deployed, it complicates the jobs of the crews when they progress from one type to the next. It places extra burdens on progression training tutorials.

Airbus has really gone a long way to implement a high degree of cockpit commonality. Ridiculed at the beginning by pilots, by experts, by technicians, but also by the competition, Airbus standardized cockpits and achieved a very big economic advantage and a safety advantage as well. When a pilot enters a new aircraft, the standardized cockpit means he does not have to be retrained.

Q: What are some applications of Real World Awareness in air travel?

A: There are plans in the cargo area, for example, where containers with precious items need to be tracked.

The RFID technology could be of great help, for instance, in luggage tracking and identification systems. Hardly any day goes by without security delays. It's a familiar announcement when a captain has to tell his passengers: "Ladies and gentlemen, unfortunately, we cannot take off because we must unload a suitcase." Oftentimes, they want to unload one that is not on board at all. If you had some way to query the luggage in the hold and tell what was there, much time could be saved. But this is just the beginning. There are vast opportunities.

Q: How would you like to see the sort of Real World Awareness we see in aviation show up in business?

A: Today's pilot gets everything he needs in real time. That means that if the pilot makes an adjustment that might lead to an unsafe situation, the computer tells him: "Friend, I'm not doing that because, otherwise, you would get into a dangerous flight attitude."

Why can't we have the same sort of standards in business? It still takes too long for me until we get our weekly results, our monthly results, and then the result at the end of the year. We are not fast enough yet. We should really be able to relate results to decisions that have been made beforehand.

The analysis of the information also seems to start from scratch too often. Why can't we define a business envelope at all different levels of the company and have automated analysis tell us when we are nearing or exceeding limits – a kind of early warning device?

TABLE 1.1 STAGES OF EVOLUTION FOR REAL WORLD AWARENESS IN AVIATION AND BUSINESS

Stage of Evolution	Aviation	Business
Seat of the pants	"Rear end" serves as sensor.	Intuition and guesswork are used.
Better information through Real World Awareness	Basic instruments are used.	Real-time data replaces batch processing and nightly or weekly updates.
Process improvements through automated response and analysis	Navigation is automated.	Key processes, like vendor-managed inventory, are automated.
Business innovation through networked intelligent systems	Autopilot flies and lands the plane, and ground proximity warning systems provide alerts in case of problems.	Adaptive business networks coordinate activity across value chains; disruptive processes, like predictive maintenance, change paradigm.

The challenge for applying Real World Awareness to business is that millions, if not billions, of points of data could be collected. But not all information is created equal. Executives seeking to take advantage of Real World Awareness must first focus their attention on collecting the information that will make the most difference. This information can then improve the performance of existing business systems, pave the way for incremental improvements in process, and eventually lead to innovations that may reshape a company and open up entirely new lines of business.

Later chapters will answer the question "Why should any company seek to improve its business through Real World Awareness?" The rest of this chapter examines the three stages of evolution and their implications and shows you how the messages for business that surfaced in our analysis of aviation relate to each stage.

BETTER INFORMATION

Huge areas of a business that were formerly in the dark can now be monitored precisely. The following exercise can help determine where to start with information gathering:

1. On a blank piece of paper, write *Seat of the Pants* in one column. List in that column every area in which you must make decisions based on inadequate, inaccurate, or stale information.
2. In the second column, write *Real World Awareness,* and describe for each Seat of the Pants area the information needed to make better decisions.
3. Sort the list by priority of the business issues addressed.

This exercise is one oversimplified way to generate a to-do list for the initial instrumentation of your business. Many companies find that this burden has been taken off their hands because important partners have forced the issue and mandated the use of Real World Awareness, in most cases, RFID technology.

Competition is also forcing the issue in many industries. Most leading firms in manufacturing, retail, and high-tech already have heavy investments in Real World Awareness technologies.

The driver for these investments is the absolute faith these companies have that more information will lead to business value. In other words, they have heard the first message we discussed earlier ("Flying by the seat of your pants is acceptable only when you have no other option") and decided that they have another option. The first phase in which better information is required begins at most companies with a similar leap of faith.

What happens is that an infrastructure investment is required that then yields real-time information—about inventory at a warehouse, for example, or work-in-progress information on a shop floor—that is used to replace stale information in existing ERP or SCM business systems.

But this stage is quickly followed by attempts to address three more messages, contained in this stage for aviation:

- As the number of sources and quantity of information grow, analysis must be automated.
- More sensors demand better connectivity.
- More sensors make standardization even more vital.

In creating the infrastructure to gather more information, companies generally expand their use of wireless networks, develop or adopt interoperability standards, and enhance their capabilities for both ad hoc and automated analysis of the increased trove of information.

Soon, this information leads to new ideas about how to improve operations.

PROCESS IMPROVEMENTS

More information almost invariably leads to opportunities to optimize current business processes and expand automation, if those opportunities are supported by technology that analyzes and manages the data to help execute a timely response. A steady stream of updates from a warehouse management system that tracks inventory at a retailer's distribution center allows the manufacturer of products to automatically send shipments to keep inventory at proper levels. When this technique, called *vendor-managed inventory,* is working properly, it serves to reduce inventory levels *and* reduce stock-outs, in which a retailer runs out of a product that consumers want to buy.

Real World Awareness allows an end-to-end view of business processes that span across companies. Applications of Real World Awareness increase security and reduce theft. Reaction times to events become faster. Consumer behavior can be closely tracked and used to transform value chains into a demand-pull model rather than a supply-push inventory model. Such developments echo the following message from aviation: "More information at first improves the current way of doing things and then creates completely new ones."

Gradually, the level of automation in business relationships increases to the point where the following message from aviation applies: "Increased automation leads to management by exception and insight." This statement means that processes that people had control of are now running without human intervention, which sometimes creates a situation that confirms the message "Some changes may not be well received."

BUSINESS INNOVATION

Although efficiency and optimization are a key focus, at every stage of this transformation new business opportunities may arise. Better information and process improvements often are part of the process of developing a deeper understanding of a business and can easily lead to new offers to customers or different ways of pricing. But the most dramatic form of transformation comes from the last message from aviation: "Intelligent automated parts of a system can lead to dramatic new levels of efficiency, coordination, and collaboration in an adaptive business network."

In retailing, Metro Group has pioneered the introduction of Real World Awareness into stores to change the shopping experience to make it more dynamic. Products with RFID tags can interact with shelves, shopping carts, information displays, cash registers, and scales all equipped with readers that can sense the products and react. Procter & Gamble is using Real World Awareness technology to pursue its vision of a consumer-driven supply network, in which timely information is used to pull, in effect, products through the supply chain rather than use planning algorithms to push products to where forecasts say that they might be needed. In aircraft maintenance, sensors on jet engines are being used to sense the operation of the engines and predict when they need maintenance or replacement of parts.

Although better information and process improvements can bring great value, business innovation based on the power of Real World Awareness is the path to the largest benefits. These examples are only a sample of those explained in more detail in later chapters of how more information, advanced analysis, and intelligent components can change the paradigm for a business. Figure 1.5 shows you the broad structure of how Real World Awareness applies to business.

Two more topics will complete our survey of Real World Awareness in business: the architectural elements that are part of the solution and the management challenges involved in making everything work.

Architectural Elements

Just as the basic structure of planes and the processes used to fly them changed as more Real World Awareness was introduced, in business the design of business processes and the relationships between companies and IT infrastructure will also change dramatically.

Business systems that were built to use stale information or data describing fuzzy aggregates will have to change. New algorithms are required in order to take advantage of more timely and accurate data. A richer virtual model of the world must be created. Business systems are supplemented by automation distributed in intelligent components that gather information, report on significant events, and otherwise act independently.

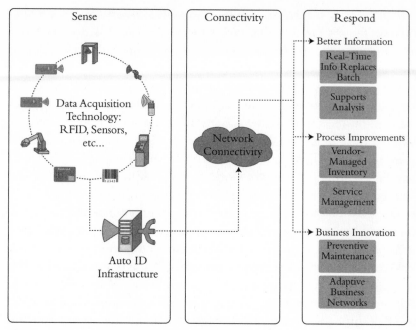

Figure 1.5. Real World Awareness in Business

For business processes, automation extends to cover a larger portion of a business and the business itself may be participating in a large adaptive business network.

Management Challenges

Managing the application of Real World Awareness requires a constant search for many different sorts of threats to success. One of the barriers covered in Chapter 5, which covers issues related to the implementation of Real World Awareness, is the difficulty of managing change. Pilots initially were skeptical and sometimes rejected certain advances, and it is likely that many companies will face similar resistance as a cultural shift toward management by exception and insight takes place. New skills will be required, and older skills may indeed be replaced. Executives will also likely have to beef up their analytical abilities to understand and take advantage of the abundance of information about previously unmonitored activities.

Perhaps the largest management challenge is one of vision. Some companies will be the leaders in Real World Awareness and will become the masters of new forms of adaptive business networks and other new structures for industries. Other companies will succeed as master participants in networks and become preferred providers. Still other companies will scramble to catch up or even survive.

This chapter has taken a look at how Real World Awareness has changed the fundamental nature of aviation and how it is beginning to transform business in the same way. Aviation has become safer, cheaper, and more reliable as Real World Awareness has been implemented. Businesses that are still flying by the seat of their pants now have a choice to make. Will continuing this way work, or will a clearer picture of the real world be required in order to survive and win?

The goal of this book is simple: to help executives and managers understand how to think about the changes coming down the road. This is not primarily a book about RFID technology—it is a book about Real World Awareness. We are just at the beginning of innovation in this area, and new technology will arrive at a steady pace. What will be more stable are the stages of change that Real World Awareness will follow. We hope that this book presents those stages in a useful way so that you and your company can thrive.

2

How Real World Awareness Will Change Your Business

Nature can tell us about how large groups sense and respond to information. A herd of zebras on the African plain munches leisurely on the grass. As the wind changes direction, one zebra smells something and sees a hint that a predator is approaching. The zebra makes a warning sound and moves away from the threat. The rest of the herd is put on alert by this behavior and, before long, all the zebras have propagated the warning and are galloping away. This pattern relies not only on information being sensed, but also on an internal model in the zebras' brains of how to instantly use that information and respond in a way that improves their chances of survival.

The question facing many companies interested in becoming more efficient is "How can we sense the important events in the real world and quickly respond with behavior that leads to success?" Real World Awareness provides the means to sense the real world, but then creates a new burden by generating mounds of raw data. To react effectively, companies must sift through that mountain of data to find the meaningful nuggets, use that information to understand a detailed model of the real world, and then respond with appropriate action.

The fact is that if most companies could wave a magic wand and instantly have RFID tags on all their products along with readers to report on their locations at key points, the amount of information would be overwhelming and crafting a response would be almost impossible.

If Real World Awareness means an explosion in nerve endings, one response must be for companies to build a bigger brain, or even multiple connected brains, to understand the vast quantity of information and craft a response. This chapter asks two simple questions:

- Will any patterns occur in the way business reacts to the abundance of accurate information that Real World Awareness can provide?
- How can companies prepare?

It is still early, but the shape of an answer is coming into focus. To explore this question, this chapter takes a close look at the retailing and consumer products industries that have been pioneers in applying Real World Awareness. The concepts of the adaptive business network and preventive maintenance are proposed as new paradigms that Real World Awareness will accelerate and enable. The chapter then revisits the three categories of responses introduced in Chapter 1 and examines how companies can prepare for each one. The chapter concludes by examining the range of applications of Real World Awareness. But, first, the time has come for a precise definition of Real World Awareness.

REAL WORLD AWARENESS DEFINED

As Chapter 1 demonstrated by using the history of aviation, Real World Awareness is a general sense that includes almost any industry. However, the definition for the purpose of this book is as follows:

Real World Awareness is the ability to sense information in real-time from people, IT sources, and physical objects—by using technologies like RFID and sensors—and then to respond quickly and effectively.

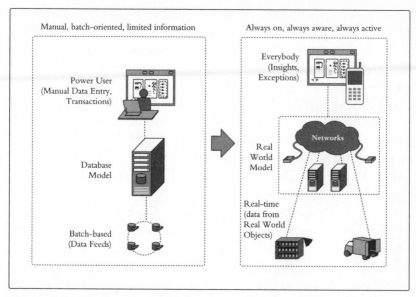

Figure 2.1. The Fundamental Effect of Real World Awareness

Like the Internet, Real World Awareness is a general technique with broad implications. So far, the changes spurred by Real World Awareness have been grouped into three categories: those that take advantage of better information, improve processes, or enable business innovation. As Figure 2.1 illustrates, the general idea is that as the volume of data increases, responses can be evaluated faster and even automated in many situations. Batch processes become real-time. A small amount of information known only to power users becomes a wealth of insights known to everyone in the company. To examine the specific effect of Real World Awareness, we now turn our attention to consumer products.

THE CHALLENGE IN CONSUMER PRODUCTS

The intricate relationship between retailers and manufacturers of consumer packaged goods (CPG) provide an excellent forum in which to expand your understanding of Real World Awareness. The CPG industry has one of the most advanced and responsive supply chains but has reached a limit of sorts in its progress. Real World Awareness may play a key role in overcoming this limit. Other industries—such as aerospace, defense, high-tech, and agriculture—are facing similar challenges.

The world of consumer products has always been keenly competitive. The goal sought by most retailers and supported by manufacturers is to make sure that when consumers walk into a store, the products they want are on the shelves and at the lowest possible prices. At Procter & Gamble, the company seeks to maximize its potential to satisfy the customer at the two moments of truth: when the customer buys the product and when he uses the product.

But the bar has continually been raised. Customers are less brand loyal than they were before. They have less time for shopping. When customers approach a shelf, if their chosen products are not there, one study showed that they will choose another brand 20 percent of the time and perhaps stay with the new brand, and, 30 percent of the time, will go to a different retailer. In addition, the Internet provides voluminous information for comparing products and has become a platform for delivering an ever-increasing stream of targeted promotions. More products are introduced faster, further expanding the choices available to consumers.

Manufacturers must therefore respond to changes faster than ever. New products must be introduced quickly to enhance a lead or to catch up with competitors. The number of new product introductions has exploded. It is crucial for companies to respond if a new product by a competitor is successful. Lots of new product introductions fail. Finding out that a product is failing and killing it quickly is important, which means that getting early information from the store is vital. If a product or promotion is successful and consumers are flocking to stores to buy certain products, the manufacturer must make sure that the products are there or else the opportunity is wasted.

One response to this problem is to pack the distribution centers, stores, and shelves with inventory. But all that inventory represents an investment that must be financed.

Another response is to improve planning and demand forecasting. For some time, manufacturers and retailers have increasingly shared information about manufacturing schedules, promotions, and point-of-sale data, in an attempt to improve predictions about demand.

To truly solve the problems facing consumer products, a new paradigm must be created.

MOVING FROM PUSH TO PULL

What has worked to keep shelves stocked and inventory levels as low as possible is closer cooperation between retailers and manufacturers. Retailers, who are the masters of the channel, and manufacturers—usually the masters of the brand—have created a consumer-driven supply network, which is one form of an adaptive business network. The basic idea behind this network is that inventory is pulled from the supply chain by consumer activity, not pushed to it from inventory.

The aim is to streamline processes so that when inventory levels start to drop in the store, the manufacturer is instantly notified and moves to replenish the inventory. One process, known as *vendor-managed inventory,* or *VMI,* starts to approach a pull model. Under VMI, the manual process of generating a purchase order for new products is eliminated. Data describing inventory levels at a retailer's distribution center is sent regularly to the manufacturer, which decides when replenishment should take place. The replenishment orders are generated automatically. Further optimizations of this process involve process innovations such as cross-docking, in which inventory is moved from the manufacturer's truck to the retailer's truck at the distribution center without ever having to be stored in the warehouse.

Although VMI is a clear step forward, it is just the first glimmer of a true consumer-driven supply network, which must handle several complexities of the consumer products industry in order to come to life.

REQUIREMENTS FOR A CONSUMER-DRIVEN SUPPLY NETWORK

The primary reason that VMI is not a true consumer-driven supply network is that the flow of information is one step away from the consumer. It is true that products move out of the distribution center because of demand from consumers, but a true consumer-driven supply network has an unbroken chain of information from the factory to the shelf at the retailer.

If sales are spiking because of a product promotion or an unexpected event, like the serendipitous mention of a product in a major motion picture, the demand planners at the factory should be aware of it *as it is happening* so that they can adjust their schedules. Several problems stand in the way of realizing this vision, however.

First, companies are increasingly global in their operations. So, for a true consumer-driven supply network to be in place, distribution centers and retailers around the globe have to be reporting sales in real-time. What is important is not that every deli and convenience store needs to be reporting demand in this way, but rather that enough of them do in order to accurately model total demand in hourly and daily cycles, not in weekly and monthly cycles.

After better information is in place, demand for a product becomes readable on the corporate dashboard, like an altimeter in a cockpit that indicates the height of an aircraft. Then, production can be more tightly mapped to demand and inventories can be further reduced. Products will move faster from the factory through the supply chain to the distribution centers and stores. Products will be pulled from the consumer in response to demand rather than be pushed from the manufacturer to the retailer based on a planning cycle and demand forecast.

The increased velocity of products has advantages, but it has drawbacks as well. At the end of the day, a consumer-driven or any other type of trading network is only as strong as its weakest link. Interruptions in the supply chain because of weather, political events, or transportation failures mean that there is less inventory to make up any shortages and that shelves may empty quicker. Of course, they should be able to be replenished quickly after the problem has been overcome. Markets have already shown they are very sensitive to supply chain problems and punish companies who do not perform at a high level. A study from the Georgia Institute of Technology showed that companies that reported supply chain problems lost 9 to 20 percent of their market capitalization over six months.

But, if you look at even the most advanced operations and relationships between retailers and manufacturers, it becomes clear that the business relationships to support the consumer-driven supply network are not yet in place. Right now, VMI means that inventory at the distribution center is visible to some extent to the manufacturer through the steady stream of inventory data used to automatically generate orders. But the in-store inventory, the product on the shelf, and the real-time rate of sales are kept within the boundaries of the retailer.

The Procter & Gamble Company, which had $51.4 billion in sales in 2003, is the largest U.S. maker of household products and leads the sector in market share with 16 billion–dollar brands. Most P&G products fall into three categories: global beauty care; global health, baby, and family care; and global household care. P&G also makes pet food and produces soap operas. P&G has nearly 110,000 employees in almost 80 countries worldwide.

R. Keith Harrison, Jr.

Global Product Supply Officer

The Procter & Gamble Company

R. Keith Harrison, Jr., was born in 1946 in Ohio. After receiving a BS degree in mechanical engineering from Duke University, he joined Procter & Gamble in 1970. He held a series of positions at Procter & Gamble divisions around the world before becoming Global Product Supply Officer in 2001. Keith is in charge of keeping the Procter & Gamble supply chain up to the challenge of manufacturing and supplying products in an increasingly competitive retailing environment in which manufacturers must become more adaptive to keep pace.

Q&A with R. Keith Harrison, Jr., Global Product Supply Officer, The Procter & Gamble Company

Q: What is the outlook for Procter & Gamble's manufacturing and supply chain systems?

A: We are clearly seeing a dramatic increase in the pace of our innovation and the complexity of the supply chains and the distribution systems. We are now looking at six or seven different kinds of retailing channels and trying to respond to the needs of each and in a very differentiated way. The challenge is to find ways to accommodate this higher-throughput, more innovative, more customized, and more differentiated world. It is obvious to us that the old kind of forecast-driven, mass-volume kind of world, focused on simplification and standardization, that we operated in through a lot of the '90s is not going to be relevant or acceptable for the world that we're moving into now.

Q: What pressures are driving changes at consumer products manufacturers?

A: Rapid change is coming from two fronts: the accelerating pace of innovation and the move to a demand-driven paradigm. Where we might have had one innovation a year in our fabric-care business, those days are gone. The pace of innovation, the innovation drumbeat, is picking up tempo. To become more demand driven, the manufacturing base is going to have to be flexible, highly skilled, and responsive. Otherwise, you are going to be carrying excess capacity or excess inventory.

Q: How will you adapt?

A: We want manufacturing driven by actual consumer demand, not forecast. We want to be producing what is selling, not what is forecast to sell. We have been working to transform our supply chains into a real-time information system that is linked end to end, is much more responsive, and is much more capable of managing complexity and differentiation. To get this done, we must be more integrated with our supply base so that our suppliers are getting the same benefits of this integration and streamlining that we are.

We want to get to the point where point-of-sale data is available on a real-time basis in our manufacturing sites and use it to drive our planning processes. We want to revise our plans as we see volume changes occur that might be meaningful.

Q: What challenges will you face in making this transformation?

A: There's a culture revolution going on. Our entire organization must understand and become excited about the idea of changing to meet new consumer demand. The old mindset, which has been in place for decades, rewarded and valued long runs of high-volume, fairly standardized products that achieved efficiency at scale. The challenge is achieving that efficency with the sort of shorter runs required to be responsive to real-time demand. Cost pressures haven't gone away, so we need to find ways to handle this differentiation and complexity better than we've ever done before.

In the absence of a flexible, responsive, highly skilled, very capable manufacturing organization, you're going to carry either a lot of excess capacity, which will drag down your cost, or you're going to carry a lot of inventory, which is going to soak up cash. Adaptive manufacturing capability is going to be key. Your logistics capability, your ability to deal effectively with your customers, your ability to integrate your suppliers, and the ability to understand time and losses across the entire supply chain are going to be the areas of growing importance and growing focus as we move forward in building out this consumer-driven supply network.

Q: What role will Real World Awareness play in making the supply chain more adaptive?

A: We want to integrate Real World Awareness technology, like RFID, on materials and finished products. Then, on a real-time basis, our information systems will always know what we have, what we've made, and where it is. We have a pretty sophisticated system today of inventory management for our finished products, but I think RFID does offer the possibility of making that a bit easier to get at and a bit more real time. But the bulk of this is going to be the integration of consumer demand data into our planning processes at the plant sites, and then the integration of those plans that are generated from that with our suppliers, all in a very transparent, integrated, real-time way. With Real World Awareness from end to end, the information can be leveraged to create immense value for us and our partners.

Q: Are standards like EPCglobal and ISA S95 key to getting this right?

A: I'm a believer in standards and protocols. I have no interest in trying to gain a competitive advantage by trying to develop a unique kind of information system. I'm more interested, frankly, in standards that allow the creation of scale and lead to the ease of development of new applications and also drive costs lower. I want us to win on our ability to understand the business environment and on our execution capability. The systems that support that should be as standardized as possible. I don't know how you get there without a robust set of protocols that can be shared across industry groups.

Q: What sort of companies will win this race?

A: The winning companies down the road are going to be the ones that have the winning supply chain capabilities. I think the battle for the consumer is increasingly going to be fought and won at the store shelf. And, therefore, supply chain capability is going to be an offensive, strategic tool for companies to use to win the battle for the consumer.

At Wal-Mart, for example, within minutes of a purchase, the record of that sale has been transported to a massive data warehouse that tracks every sale in every store. This incredibly valuable corporate asset is shared sparingly with partners on a need-to-know basis. For the true consumer-driven supply network to work, some level of direct connection to this information must be established from one end of the supply chain to the other. But the business relationships to support this and many other changes that will be required are not in place or even designed. There is an incredible opportunity for innovation in new forms of business relationships.

It is no accident perhaps that the most responsive supply chain in existence at Dell, Inc., operates in a company that combines the brand master and channel master roles. Dell is a direct retailer and also the manufacturer. Its supply chain is so responsive that it has a fraction of the inventory costs of most manufacturers or retailers and has achieved the amazing feat of having a negative cash-to-cash cycle by collecting the money for its products before it must pay its suppliers. (For most manufacturers, the cash-to-cash cycle measures the delay between paying suppliers and collecting money from customers.)

The Dell structure also solves another difficult problem in the consumer-driven supply chain: accountability. The brand master—most often, the manufacturer—is blamed for a lack of quality no matter who is at fault. Are the packages on the shelves damaged? Is the product stale? Are the prices wrong? Whatever the problem, the brand suffers the blame. In a consumer-driven supply network, the brand masters will want more information and the ability to ensure quality in all forms.

The payoff for getting this right is huge: dramatically reduced inventory, the ability to respond immediately to changes in demand, a precise understanding of the customer and how well promotions work.

Crafting consumer-driven supply networks remains a huge challenge that requires sustained effort over a long period.

WHAT IS MISSING?

The vision of a consumer-driven supply network is broad and compelling, and there is really no barrier to creating such business networks. The technology is there to support the intercompany processes, and the business relationships are starting to become better understood and standardized. Early successes have generated lots of interest, and many companies are in the process of deciding whether they are better suited to be coordinators or participants in business networks.

The challenge now is to make consumer-driven supply networks work as efficiently as possible, and even the most advanced firms have a long way to go.

The missing ingredient is information. The barrier is often the cost of collecting it. The solution is Real World Awareness. The rest of this chapter shows you exactly how the gap between current practice and the fully realized vision of what can be done with Real World Awareness will be closed.

FEEDING THE GROWING HUNGER FOR INFORMATION

A close look at the process of vendor-managed inventory (see Figure 2.2) helps explain how more information enables better execution.

Vendor-managed inventory creates a leaner process, by automating a step—that of creating a purchase order—and then shifting the responsibility for keeping inventory stocked from the retailer to the manufacturer. Here's how it works:

1. The level of inventory in the retailer's warehouse management system is broadcasted regularly to the manufacturer as it is changed as products leave the distribution center for the stores.
2. The manufacturer monitors this level and automatically generates a purchase order to itself when the inventory level drops below a certain level.

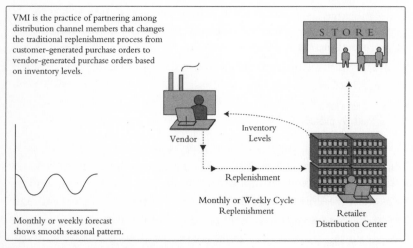

Figure 2.2. Vendor-Managed Inventory

Dell Inc. is a world leader in the manufacture of PCs, servers, and other computing devices, having revolutionized the industry with its supply-chain innovations and customer-direct sales channel. Dell's net sales have totaled $47.3 billion over the past 4 quarters, and Dell now has approximately 50,000 employees, 6 manufacturing plants, and operations worldwide.

Chairman and Founder

Dell Inc.

Michael Dell

Michael Dell is founder and chairman of Dell Inc.
Michael, born in February 1965, founded Dell Computer at the age of 19 with $1,000 and an office in a college dormitory room. Eight years later, Dell was a Fortune 500 company, and Michael himself became the youngest CEO to ever appear on the list. In addition to his duties at Dell Inc., he is an IT Governor of the World Economic Forum, serves on the Executive Committee of the International Business Council, and is a member of the U.S. Business Council.

Q&A with Michael Dell, Chairman and Founder, Dell Inc.

Q: Dell is one of the rare companies that manufactures its products and also sells directly to its customers. Why is that important?

A: We have, we think, an advantage in that we are in direct contact with our customers. And so we have real-time information about demand and the requirements, and it's not filtered through a series of distributors or dealers. We do not have to guess what our customer is going to buy, because we know. Our customers are telling us every day. We're meeting with them face to face, or they're ordering through SAP on a direct link, system to system, or it's by telephone, on the Internet, or it's the fax machine or some kind of communication. At a typical manufacturer or retailer, you might have several different layers of guessing or anticipation. RFID is one of the things that can improve that efficiency further still because it provides better information.

Q: What has been the role of technology in Dell's success?

A: We also drive a lot of planned obsolescence, a process by which we plan for the retirement of older technology platforms as we enable our business with new applications. And a significant percentage – today, it's about 62 percent of our IT staff – are working on new development projects. And we hope that by the end of this year, we'll be at 70 percent. Most organizations are flipped the opposite way – 70 percent of their resources are focused on just maintaining or sustaining existing systems, and so they never get a chance to develop new functionality or new capability they need to move the business forward.

Q: How do you identify the right moment and the right recipe for developing new capabilities?

A: It's central to what we do. Because we're in the IT business, we ourselves tend to be an early adopter. For example, Oracle started talking about the concept of clusters. We saw that was a wonderful idea and started prototyping it in our R&D labs and internal IT environments. We tend to use these things internally in our IT environment before we take them to our customers. Our IT team is leveraging much of what we're selling, and we believe that our IT structure is the best example of how to run a very large and growing enterprise.

Q: Because of that size, has Dell reached the point yet at which further acceleration of business processes has become impossible?

A: No, we don't believe that. The idea of continuous improvement is just an ongoing, never-ending focus for our company. And we're always introducing new capabilities that help us – whether it's improving the customer experience, or working on our manufacturing, or supply chain, or Dell.com. We like projects that are either 3, or 6, or 9 months in duration. In a project which is 2 years or 3 years in duration, what we find is that by the time it's done, the requirements, and our business, have totally changed. So we try to have smaller, incremental improvements that are gradual as opposed to having massive, so-called "boil-the-ocean" projects.

Q: RFID is one of these new technologies currently being tested and evaluated. Are you evaluating its riskiness, or have you embraced it already?

A: That's one technology that certainly has a significant impact for resellers. We're using it already in our manufacturing environment. It started at the pallet, but eventually it will go to the actual device. And any time you can improve the quality of information, the opportunity for business process improvement follows naturally. In a typical manufacturer or retailer, a lot of times they don't know what the demand is and how it is trending, so they just have to put the stock out here and hope the demand is there.

Under this system, the manufacturer is responsible for deciding at what level of inventory replenishment will take place.

The beauty of vendor-managed inventory is that one process step has been removed, that of manually generating a purchase order. Instead, the reported level of inventory is used as the basis for the automatic process. The manufacturer can synchronize inventory levels with promotions, base forecasting on more complete data, and reduce purchase order errors, which reduces stock-outs (the depletion of inventory).

The key here is a steady stream of information from the distribution centers to the manufacturer. The question is "How is that information collected?" Without Real World Awareness technologies, some sort of manual process, perhaps assisted by bar code readers, must keep inventory levels up to date. As inventory moves into the warehouse from the manufacturer, it must be tracked. As it moves out to the retail stores, it must be tracked. The more manual the process, the more labor is required, the staler the data, and the more errors will occur. For process innovations like cross-docking, the inventory moves faster, from truck to truck, and the window to collect data becomes smaller.

Vendor-managed inventory works well in today's environment, but imagine what would happen if the same techniques were applied to the fully realized vision of the consumer-driven supply network without the help of Real World Awareness technologies. The number of data-collection points would explode and become unaffordable. Armies of employees with bar code readers would have to scan products as they moved from the distribution centers to the stores and then to the shelves. The cost would be astronomical, and the data would be prone to error and omission. A more responsive replenishment process that could be part of a consumer-driven supply network is shown in Figure 2.3.

REPLACING INVENTORY WITH INFORMATION

The solution is Real World Awareness, which, through a variety of techniques, automates the accurate collection of massive quantities of data from many distributed locations. The unaffordable army of employees equipped with bar code readers is replaced with RFID tags on products and readers at important locations.

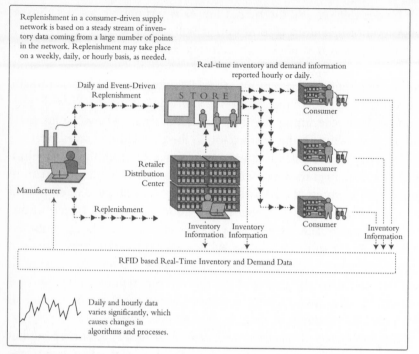

Replenishment in a consumer-driven supply network is based on a steady stream of inventory data coming from a large number of points in the network. Replenishment may take place on a weekly, daily, or hourly basis, as needed.

Real-time inventory and demand information reported hourly or daily.

Daily and Event-Driven Replenishment

STORE

Consumer

Manufacturer

Retailer Distribution Center

Consumer

Replenishment

Inventory Information Inventory Information

Consumer

Inventory Information

RFID based Real-Time Inventory and Demand Data

Daily and hourly data varies significantly, which causes changes in algorithms and processes.

Figure 2.3. Replenishment in a Consumer-Driven Supply Network

For the consumer-driven supply network, the pallets and products may be tagged and tracked by readers at many points in the process, including while these tasks take place:

1. Load at the manufacturing plant.
2. Deliver to distribution centers.
3. Load again on retailers' trucks.
4. Deliver to stores.

The METRO Group Future Store pilot, described in Chapter 3, extends this concept further and tracks inventory on the shelf, leveraging so-called "smart" shelves.

The abundance of information that is collected can be used to reduce inventory dramatically. You can easily imagine distribution centers being bypassed and shipments arriving directly at the stores from the manufacturers, and new innovations such as daily or even hourly replenishment cycles, or perhaps a structure that collapses the factory and distribution center into a single facility.

All these potential innovations are made possible by real-time information about the state of inventory all along the supply chain, which can lower inventory by speeding up the movement of the products from manufacturer to retailer. A faster supply chain reacts more to real-time signals of consumer demand and is less dependent on the quality of the short-term forecasts.

Of course, the supply chain is just the beginning. Tags can be used in many ways and in many contexts, such as providing a running total of items tossed in a shopping basket or displaying product information at kiosks. Perhaps DVDs can be waved in front of a screen that plays a sample of the songs or a trailer to the movie. The potential for tracking items is vast and just beginning to be explored.

Other potential benefits from Real World Awareness include reducing inventory shrinkage caused by theft and ensuring the quality of products by using the sort of advanced sensors for temperature or shock that we discuss in Chapter 3, which focuses on technology.

All these benefits are based on Real World Awareness techniques, such as RFID, which reduces the wholesale cost of acquiring information.

Our analysis now takes a step back from retailing and consumer products and looks at some of the general patterns of change that have started to emerge.

RECOGNIZING EMERGING PATTERNS OF CHANGE

Real World Awareness "turns the lights on," by providing reams of data and then enabling automation and new ways of doing business. So far, two general structures have started to be repeated in many different industries: Preventive maintenance and the adaptive business network.

Preventive maintenance is the practice of turning the traditional paradigm for maintenance completely around. In normal practice, equipment is put into the field for a specified period and then brought in for maintenance after it breaks or after a specified time or usage limit has run out. As it turns out, even with today's advanced manufacturing techniques, each machine is a bit different.

What preventive maintenance does is create unique definitions of what is normal for each machine being maintained, by using anywhere from 10 to several dozen measurements. Normal ranges for these measurements are established by monitoring the machine while in use. Ranges are then tuned to the behavior of a particular machine, which allows the maintenance paradigm to be changed to one in which machines are scheduled for maintenance when they start to behave differently from their normal ranges. Technicians are alerted to problems before equipment has failed. Equipment can operate without interruption beyond the historical averages if everything appears to be normal. The result that has been confirmed in maintenance of jet engines, factories, and other fields is that unscheduled interruptions are minimized and maintenance costs drop sharply because everything can be planned better.

Perhaps a more general and widely applicable pattern is the adaptive business network, a general theory of how business can be made more responsive that can be greatly enabled and accelerated by Real World Awareness. [*Editor's note:* Claus Heinrich introduced the concept of the adaptive business network in his last book, *Adapt or Die,* published by John Wiley & Sons, Inc.] The consumer-driven supply network is one example of an adaptive business network.

The concept of the adaptive business network was created in response to the needs of businesses to find a new way of operating that gives them the flexibility to respond quickly to unexpected changes. It is also a paradigm that is made significantly more powerful when coupled with real world awareness. By linking companies through standard business processes and common technology, the adaptive business network allows them to work together as a loose network of partners. By cooperating with other businesses in the network, each company can respond more swiftly to changing market conditions than it could on its own. Companies within the network remain autonomous, but are able to leverage the networks' cumulative ability to do the following:

- Plan and anticipate demand and supply.
- Execute plans efficiently and effectively.
- Sense events that affect the plans as soon as those events occur and then analyze their impact.
- Respond to and learn from ever-changing business conditions. The relationships between these concepts are shown in Figure 2.4.

The adaptive business network is designed to help businesses quickly respond to changing market conditions by capitalizing on the strengths of operating units within the company and trading partners outside the company. It is a model based on mutual goal setting, not limited to the traditional buyer-seller relationship that now exists between most trading partners.

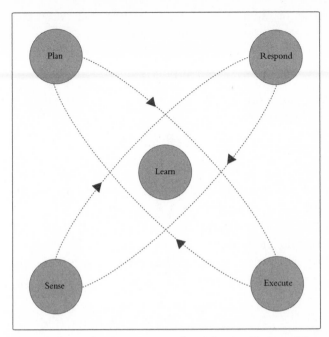

Figure 2.4. The Dimensions of an Adaptive Business Network

Companies within the adaptive business network can react quickly to changing customer demands by efficiently exchanging information, to the benefit of all participating companies. In addition, new partners can be added to the network quickly and inexpensively as market conditions change and new business opportunities arise.

Participation in an adaptive business network puts companies in a position to remain flexible, resourceful, and profitable in a constantly changing business environment. It allows businesses to meet the increasing demands of consumers who expect high-quality, personalized products designed and delivered in ever-shortening time windows, and to attract new customers and sales based on the ability to meet those changing consumer needs. The network helps reduce costs by streamlining processes to focus on what each participating company does best, and it allows all participants to collaborate dynamically with their partners to produce new and innovative products and services.

In short, the adaptive business network provides new opportunities to

- Increase profit margins
- Realize cost savings

- Accelerate cash-to-cash cycles
- Improve the effectiveness of corporate expenditures
- Capture a greater return on assets

The role that Real World Awareness plays in the adaptive business network is to turn the lights on and provide each participant in the network with the information needed to effectively play its role. To be adaptive, to react to the unexpected, to make the right decisions, each participant needs to know the state of the network, both up and downstream. In the ideal operations of the adaptive business network, information flows through the network and spurs the appropriate action from each participant to meet the needs of the customer.

These patterns are just two that we have noticed. Many more will doubtless become revealed, as companies understand how best to take advantage of the opportunities that Real World Awareness brings. The most important task for companies now is to be ready for the coming changes.

PREPARING FOR REAL WORLD AWARENESS

At the beginning of the Internet era, businesses could see only the raw potential of its functionality. Many wild ideas and theories were proposed, and some of them worked out and many of them crashed. In the end, though, the Internet affected business according to a few common patterns. Business-to-business purchasing, extranets, intranets, and e-mail all became part of the Internet-enabled corporate landscape.

We don't yet know what all those patterns will be for Real World Awareness. The way that Real World Awareness amplifies the value of adaptive business networks probably means that it will become a common pattern for highly competitive, consumer-facing industries that have complicated value chains. Preventive maintenance using predictive analytics also seems to be a powerful concept. But many others are waiting to be discovered.

Each industry is likely to figure out its own patterns, and some vendors will create killer applications and the next Amazon, eBay, or Yahoo will be born as companies learn to leverage the power of Real World Awareness. At this stage, the question is "How can a company get ready for all this?"

This book argues that the best way is to "know thyself." The more you understand your operations, the easier it is to design and implement Real World Awareness when the time is right.

STAGES OF EVOLUTION

The stages that most companies pass through as they grapple with the arrival of Real World Awareness involves first gaining access to and taking advantage of better information, using that information to improve process steps, and, finally, creating entirely new innovative ways of working. In this section, we examine how each of these stages will affect most businesses.

Better Information

Real World Awareness provides better information in many forms. Whereas previously information arrived only sporadically and inaccurately, Real World Awareness substitutes timely and real-time data. Whereas data was aggregated, Real World Awareness can provide more granularity. Whereas practical barriers existed to collecting information, Real World Awareness can overcome physical and logistical limitations.

At first, this information can make a difference by simply replacing batch information or data that may be stale or inaccurate. A warehouse management system that has real-time data is more useful than one that has batch data updated once a day. As time goes on, companies can provide the information to their partners, or analyze it to improve their understanding of their business.

In this first stage, skill is built in data-collection and aggregation techniques. Real World Awareness will likely lead to an expansion of data warehouse, data mining, Online Analytical Processing (OLAP) techniques, and other technologies and analytical techniques used to extract meaning from the increased amount of data.

The sort of questions that companies may find useful to ask in advance include the ones in this list:

- Where will real-time information help?
- What important information is unavailable?
- What information could suppliers and partners find valuable?
- How will processes have to be changed?
- What analytical capabilities will be needed?
- What new metrics can be monitored?

Process Improvements

It does not take long for better information to lead to ideas about how to improve or automate processes. In fact, in many cases, using better information requires new processes and algorithms. For example, moving from weekly and monthly replenishment cycles based on forecasts and sparse demand data to a daily cycle and event-driven replenishment based on daily or hourly data almost certainly means that a new process and new algorithm will be needed.

Increased information also leads to more opportunities for automation of the sort shown in Figure 2.5. Vendor-managed inventory allowed the automation of inventory management based on a steady flow of data. Many such opportunities present themselves as the number of data points about a process increases. One such example is the automatic generation of advanced shipping notifications based on RFID data. If pallets are loaded with RFID tagged boxes and then the pallets are tracked as they are loaded onto trucks, it should be possible to send a completely accurate advanced shipping notification to the receiving company.

Reading boxes without having to open them may lead to other sorts of process improvements. Tagging components that are assembled into modular products, such as computers, offers new avenues for quality control.

The identification numbers on the tags can be used as indexes to all sorts of data, kept on the servers, that can lead to increased automation and many different levels. Data can start to move around on tags and be distributed in other ways, which opens the door for new ways of working.

The following sorts of questions lead to the discovery of areas in which Real World Awareness may be able to extend automation or improve processes:

- Where are errors and delays?
- Where do the processes need to speed up?
- How could processes change to take advantage of better information?
- What automation does better information make possible?
- What processes have bottlenecks?
- What can be done to improve partner and supplier processes?
- What type of data should be stored on the tag for use by other parties?
- What is the right distributed data architecture?

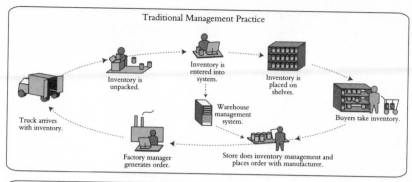

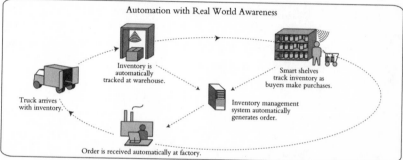

Figure 2.5. Increasing Automation with Real World Awareness

Business Innovation

The business innovations that spring from Real World Awareness will probably have the largest and longest-term impact. Entire industries will be reshaped as the cumulative effect of better information and process improvements make new forms of organization possible. Early examples are the way that Dell has combined the brand and channel master roles to great effect. Aircraft maintenance is being revolutionized by preventive maintenance techniques that closely monitor the individual profiles of each jet engine and predict when maintenance is required before problems occur. Large organizations, like Wal-Mart and the U.S. Department of Defense, are mandating the use of RFID tags to improve efficiency from one end of the supply chain to the other.

Clearly, these transformations will produce both winners and losers. The question that each company must ask itself is "Will our company be the orchestrator or a participant in the new world?" In many cases, companies will be both.

These operational innovations are likely to cause wrenching cultural change. As Keith Harrison, of Procter & Gamble, pointed out in his interview in this chapter, long-held assumptions about the goal of manufacturing (the optimization of costs over long runs), will now have to change to a more adaptive form of manufacturing in which costs are held as low as possible as bursts of products are created. Assumptions will be changed in the same dramatic fashion in many industries.

Increased automation will also lead to the rise of the "management by exception and insight" paradigm, as shown in Figure 2.6.

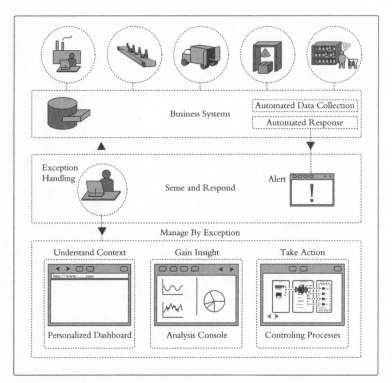

Figure 2.6. Management by Exception and Insight

Management by exception and insight means that organizations retool from being hands-on actors who guide the execution of processes to quick-response teams who react to alerts raised by automated systems. After those alerts are raised, the team must use the information that is available from all sources to determine a correct response. Few observers expect this paradigm shift to be easy.

Business innovation is the most challenging aspect of Real World Awareness that will lead to significant cultural change and deep questions of corporate identity. Here are some questions that companies will ask along the way:

- What are key pain points?
- How can I reinvent my business to address them?
- What role will my company play as the industry changes?
- How much risk should my company take?
- How will management attitudes need to adapt?
- What will management by exception mean for my company?
- What analytical tools will be needed to manage exceptions?

Two other aspects of Real World Awareness will run through all the evolutionary stages (better information, process improvements, business innovation) that are analyzed in this chapter: architectural and management changes. Real World Awareness will put pressure on several different layers of the current IT infrastructure. Better information and process improvements will require that data models and architectures change. Business innovations will create a variety of management challenges as well.

Architectural Changes

The architectural changes that take place as the use of Real World Awareness expands will be driven by the requirement to manage and analyze a much larger set of data and then use it to automate processes.

Richer Models

Real World Awareness will require a company-wide enhancement of data models. Today's business systems were built to manage yesterday's granularity and volume of data. Now that much more is known about the world, business systems will need a richer model to track and make sense of all the data. In fact, the model of the world in many applications will be so detailed that it will in many ways no longer be a model; it will be a mirror of the world that can tell where hundreds of thousands or millions of objects are at any given minute.

The shop floor or warehouse now can be mapped more thoroughly with a model that can easily be kept up-to-date. The location of pallets and products in the warehouse can be precisely tracked, and the order in which they should be loaded on to trucks can be planned more precisely.

The model must be carefully designed to support the processes and responses involved in the related business scenarios. This model must not be frozen in stone, but can evolve as innovations are discovered and learning takes place.

Distributed Intelligence

Real World Awareness systems are distributed by necessity. They cannot work any other way. RFID readers are relatively simple devices that receive huge amounts of raw data. A layer named Automatic Identification (Auto-ID) Infrastructure makes sense of that data, stores it, massages it, and links it to objects such as purchase orders in business systems. As tagged items move around, readers and supporting systems must go wherever they are needed to keep track. System architects have a major challenge in breaking up business systems into the right layers so that functionality can be distributed all over the globe. In the future, embedded systems as part of RFID tags, RFID readers, appliances, and machines will drive the distribution of computing power in the network and will perform local business logic that leverages the available local data. For example, a shelf in a supermarket will become "smart" via embedded RFID readers and a system that monitors the item level inventory on the shelf and triggers a signal to a central store system if store personnel should replenish on the shelf from inventory in the backroom.

Connectivity

Wireless networks are rapidly becoming part of the landscape of most corporations. For Real World Awareness, wireless networks are crucial to allowing devices with Real World Awareness to be spread over factory floors as well as unwired or difficult-to-wire environments at low cost. A high-speed wireless network is likely to be a precursor to many Real World Awareness implementations.

Standardization

Given the networked and distributed nature of Real World Awareness systems, it would be almost impossible to imagine its widespread adoption without some accepted standards. EPCglobal, which was the offspring of the MIT Auto-ID Center, has provided that standard so that companies can rest assured that tags written by one company can be read by another. The EPC standard also defines ways to look for information related to a tag.

Standardization, of course, creates a network effect across partners. As more companies use EPCglobal tags, even more companies can benefit from information. Standards also allow systems originally developed for closed-loop operation inside a company to easily become part of an open-loop architecture that spans company boundaries.

Automated Analysis and Response

The volume and granularity of data driven by Real World Awareness create a new opportunity for business insights, but also drive the need for advanced techniques to automate the analysis of and response to such data.

Automated responses may have different facets, from business process management systems triggering workflows to rules-based systems for automated decision making in certain situations to new mathematical algorithms providing decision proposals in certain situations. The richer model and the increase in data can support more automation. Software agents locally monitor and control process execution at any time.

BUSINESS CHALLENGES

Real World Awareness will have a disruptive effect on the structure of most organizations. Better information, process improvements, and business innovations will bring with them several sorts of challenges.

Responding to Increased Demands for Analytical Ability

The common practice now for data analysis is reporting. Even the best financial and controlling professionals around the globe can digest only a certain number of data points in a given report. Most human brains find it difficult to identify correlations or patterns. Real World Awareness requires a company to build its analytical capabilities, supported by business analytic technologies, like Online Analytical Processing (OLAP), data mining, and predictive analysis, as well as the ability of its users to leverage such techniques to gain the business insights needed to drive business decisions at any time.

A richer model, more data, and automated monitoring and analysis provide the basis to understand patterns coming in from the real world. When everything is running smoothly and automated processes carry on the job of directing the activity of a company, management does not have much to do. But, when something goes wrong, managers must be able to rapidly sift through data and understand what is going on in order to craft an appropriate response. Expanding the automation of analysis also expands the scope of automation. Companies are likely to consciously have to increase their capacity and level of expertise in analytical techniques such as data mining, OLAP, and related techniques.

Managing Change

Just as pilots resisted the departure of navigators from their cockpits, so will many people resist the changes that Real World Awareness will bring. Some will object to the changes in processes, and others to lost or reconfigured jobs. One of the biggest challenges in introducing any technology is understanding a company's capacity for change. Introducing Real World Awareness will provide a challenging test.

The Cultural Shift to Manage By Exception

The challenge of management will shift from handling and managing normal conditions, which will increasingly be automated, to helping a company react quickly to unforeseen events and opportunities. In a fully functional consumer-driven supply network, the velocity of inventory will increase, which means that less inventory will be on hand at all points in the value chain and less time will be available to respond to events.

Companies will have to develop a much richer set of responses, a playbook of sorts that will enable rapid responses that are anticipated and that accelerate the crafting of responses to new situations. These needs will change the way companies are organized and think about what they do, which is never easy.

In many ways, this new world of sensing and responding with Real World Awareness resembles the World Wide Web. Each nexus of information, each reader or model of a warehouse, has a rich set of information associated with it and is publishing important events for use by others. What companies need to sense and respond is a sort of analytical Google that harvests information from all the Web sites in an adaptive business network and helps make sense of it and prepare a response.

Creating a Vision and Defining an Identity

Some companies, like Wal-Mart, mandate change in an industry. Others, like Procter & Gamble, are expert participants in many forms of supply chains. Companies like Dell combine the roles of manufacturer and retailer. Each company will have to understand how to make its strengths work for it in the new world that Real World Awareness will bring. Companies that know what they are good at doing will avoid flailing and wasted effort as they adapt to new business conditions.

One result of the emphasis on the retailing and consumer products industry is that perhaps Real World Awareness has been presented too narrowly. To complete this chapter, we guard against that impression by looking at two innovative applications of Real World Awareness.

APPLICATIONS OF REAL WORLD AWARENESS

Attendees of the World Cup in Germany in 2006 will have a slightly different experience than they do at most other sporting events. Each ticket will have its own RFID chip, which will be used to control access to games, reduce fraud, and provide information and a variety of other services.

That RFID chips are small enough and cheap enough to be affordable for such a purpose is no surprise to those who have been following developments closely. Other Real World Awareness technologies, like sensors and biometrics, are progressing very fast. Most promising is the trend of combining these technologies into one entity, such as putting RFID, sensor, and computing capabilities into one Real World Awareness chip. The cost of chips has been beaten down steadily, through the inexorable march of Moore's law about the rate of increase in and falling cost of processor power, advances in miniaturization, and continuing progress in antennae technology. RFID has gotten so much attention because of mandates by Wal-Mart, the United States Department of Defense, and others that many people have a limited notion of RFID and other Real World Awareness technologies as a souped-up version of an electronic bar code.

The truth, of course, is that Real World Awareness technologies have been making dramatic advances in all sorts of directions. They are not just getting smaller, but are getting more powerful, storing more information, containing more logic and intelligence, and solving many more problems than an electronic bar code ever could. The surrounding infrastructure of wireless technology has also made great strides, which eases the burden of deploying Real World Awareness systems. The EPCglobal standard for RFID, which is based on more than a decade of work at MIT's Auto-ID center, describes an architecture not just for the tags, but also for the surrounding software and information services needed to solve the entire problem.

Here are the sort of innovations that these technological advances will enable:

- Phones tagged with RFID chips that also have RFID readers to allow advanced services, such as in-store product comparisons based on the retrieval of information linked to tags
- RFID tags included in surgical devices, such as sponges, to make sure that no such device gets left inside a patient
- Temperature sensors attached to shipments of fresh food, like fish, to ensure that refrigeration remains constant
- Sensors and adaptive headlights in cars to ensure the best light distribution in driving through curves at night
- ISA S95 standard to standardize data integration from shop floor control systems with ERP systems

RFID tags are even being embedded under the skin to control access to sensitive information or to nightclubs.

TELEMETRY IN FORMULA ONE CARS

Although all these applications are exciting and represent powerful tools for convenience, innovation, and automation, they are all rather conventional. Our last example goes beyond RFID to show how Real World Awareness is being used in the fiercely competitive field of Formula One racing. The lesson here is simply that if Real World Awareness can work at close to 200 miles per hour, it ought to be incredibly useful, and practical, at the velocities found in most warehouses and executive office buildings.

Formula One racing has always pushed the limits of automotive engineering. Many features that consumers take for granted in their own cars—disc brakes, torsion-bar suspension, and overhead camshafts—were first tried out on the Formula One circuit. Lately, Team McLaren Mercedes has taken another technological step forward, by equipping its cars with an advanced telemetry system that is quite similar in concept to what is used on spacecraft. A web of sensors throughout the cars can collect reams of real-time information about many aspects of the car's mechanical health and performance. Yet, even with the cars hurtling down straightaways at 200 m.p.h., this data reaches race engineers within milliseconds of being collected.

During a typical race, the more than 50 sensors in these sleek racing machines collect and pass along as many as 1 billion parameter values—enough data to fill almost 1,000 London telephone directories. Sensors mounted throughout the engine and chassis monitor all the important characteristics affecting handling, aerodynamic performance, and fuel. Armed with all these real-time readings, race engineers can make faster and better-informed decisions about race strategy, how to set up each car in response to rain and other changes in track conditions, and how to deal with mechanical or other problems that can now be caught almost immediately as they appear. These types of decisions can make the difference between winning and losing a race.

Moving so much data from a vehicle traveling as fast as these cars do to a garage kilometers away is no trivial undertaking. The McLaren Mercedes telemetry system must compete for radio "air time" with not only other racing teams doing the same thing but also the press' video links. And, all this telemetry must work in an area surrounded by grandstands, trees, and other obstacles. The McLaren Electronics engineers put their minds to the problem, though, and conceived, developed, and manufactured a telemetry system that exploits some of the same features on a race track that cause conventional RF (radio frequency) systems to fail. They have thereby produced telemetry that consistently delivers more than 99 percent of the data sent on all racing circuits around the world, from the streets of Monaco to the long, tree-lined straights of Monza.

Although the example of Real World Awareness in Formula One is a display of raw technology muscle, our concern is more practical. One way to look at the future of business is to examine the progression that will take place from vendor-managed inventory to a consumer-driven supply network. Although this challenge is a major one, perhaps the bigger challenge is to start to understand what new structures become possible in a world that is fully populated with Real World Awareness. Has Michael Dell discovered a rule that the brand master must also become a channel master to achieve the highest levels of efficiency, adaptability, and intimacy with the customer before and after the sale? What other rules remain to be proposed and tested in the market?

This future is still cloudy and just coming into shape. Whoever comes out on top and creates the new Dell or Wal-Mart, however, is likely to do so through a strategy based on a keen understanding and shrewd use of Real World Awareness. Will this be an accident? Most certainly not. The winners will be the companies who see the opportunity early and move quickly to take advantage of it. We hope that, for some people, the journey starts with this chapter and the information that follows.

McLaren Racing, one of the most successful Formula One racing teams in history, traces its lineage to 1963, when Bruce McLaren, a New Zealander, formed the team. The team won its first Formula One race in 1968, which began a series of wins including 11 Drivers' and 8 Constructors' World Championships. Regarded by many as the pinnacle of auto racing, Formula One is the most technologically advanced and challenging sport in the world, where the boundaries of technology are continuously pushed.

Ron Dennis

Team Principal

Team McLaren Mercedes

Ron Dennis is the Chairman and CEO of the McLaren Group and Team Principal of the Team McLaren Mercedes Formula One team. McLaren Racing, the company behind Team McLaren Mercedes was formed in 1980, following the merger of Team McLaren and Project Four, Ron's racing company. Under Ron's leadership, the team has been one of the most successful at winning World Championships and innovating technology. Ron, who was born in Woking in South East England, was honored with the appointment as a Commander of the Order of the British Empire, CBE, in the Summer of 2000.

Q&A with Ron Dennis, Team Principal, Team McLaren Mercedes

Q: Team McLaren Mercedes takes Real World Awareness into one of the most challenging and high-performance sports. How does it work?

A: Like many good engineering solutions, the concept behind the telemetry system is elegantly simple. There is a huge amount of data to send, and the operating environment is difficult and unpredictable. So, the answer is to choose a high enough frequency to provide a reasonable bandwidth (for example, microwave) and then split the data into many small pieces and send these pieces concurrently over a range of finely spaced carrier frequencies. Provided enough of the pieces arrive intact at the other end, the original data stream is re-created.

Turning this concept into reality requires some heavy-duty processing in the car and garage and high fidelity in the Radio Frequency (RF) electronics. The processing is done by top-of-the-range DSPs (Digital Signal Processors) in the transmitter and receiver. Each of these DSPs is capable of processing billions of instructions every second. The fidelity is achieved by using specially designed, and extremely linear, power amplifiers in the RF circuits.

Q: So, how does data from a sensor on the car find its way to the race engineer in the garage?

A: Firstly, the measured value from the sensor is logged in the ECU (Electronic Control Unit) and sent in a data stream to the transmitter. The DSP in the transmitter then modulates the data (assigns data to different carriers), encodes the result, and arranges the carriers into a sequence, adding pilot (reference) carriers that will be used by the receiver to help line up the data correctly at the other end. The data is then manipulated further to establish an optimum configuration for transmission.

The transmitted signal is picked up by a receiver which correlates the data (finds the start of each frame of data) and corrects for any small timing differences between itself and the transmitter. The signals are then decoded and demodulated to recover the original data stream. The data is fed into a PC in the garage, which has special software installed to gather and distribute the information around the garage. The Advanced Telemetry Linked Acquisition System (ATLAS) Software has been specially built to deal with the huge amounts of data in real-time.

The race engineer is now able to look at the output of the sensor on his PC (the so-called "battle station"). In this way, a change in a sensor value is seen in the garage milliseconds after it has happened. Indeed, you will often see a gear change on the screen before you even hear it happening on the circuit.

Q: Is the data secure?

A: Yes. The data has a unique identifier embedded in every message that is recognized only by the team's garage software. The messages are also "scrambled" prior to transmission, providing an extra level of security. There is, of course, another, quite fundamental, level of protection. The decoding of the transmitted data stream relies upon the complex processing that is done by the DSP in the receiver itself, and these receivers are available only from McLaren Electronic Systems. So, there is no chance of using a third-party receiver or scanner device to intercept the data.

Q: Is there ever perfect coverage?

A: Unfortunately not (or at least, not often). You get very close to perfection on many circuits, but some data will invariably be lost in the trees, behind the grandstands, or simply through interference from other RF traffic. However, the telemetry system has also been designed to deal with this situation. Again, the solution is beautifully simple: It exploits the available bandwidth to send the data more than once. The system also knows where on the circuit the telemetry works best, so it can wait until it is in an area of good coverage before resending data recorded in the RF "black spots."

Q: What equipment is needed?

A: On the car is a transmitter and antenna. The transmitter is the size of a small book, and the antenna is a bit like a small shark's fin, about 30 millimeters long.

In the garage is a receiver, its power supply, and a splitter box (that allows receivers for two cars to share the one antenna). These parts are about the size of a long shoebox. Behind the garage, on top of a 15-meter-long mast, is a small antenna.

The transmitter is attached via a high-speed serial communications link. The receiver is connected to the garage computer network via Ethernet.

Q: How is it licensed?

A: Every team must get permission to use specific frequencies at a race. Everyone uses telemetry, so it is vital that systems use bandwidth responsibly and efficiently to ensure that everyone fits into the available "space." The telemetry used by Team McLaren Mercedes operates at a frequency of about 1.5 GHz and uses up about 5 to 6 MHz of bandwidth.

Q: What is legal?

A: In Formula One, sending data from the cars to the garage is unrestricted. Despite a brief respite in 2002, sending data back from the garage to the control system on the car is now illegal. It is still permitted to use handshaking, provided there is no possibility of transmitting information from the garage into the ECUs on the car.

3

The Technologies of Real World Awareness

N ot so long ago, only a few people had the freedom to make telephone calls from "out on the road," regardless of their locations. Although cartoon detective Dick Tracy may have been able to talk through his wristwatch, the first mobile telephones were so bulky that it took a car to lug one around. And, in any given city, there was room on the airwaves for only a handful of mobile callers at a time.

Now, of course, hundreds of millions of people from every walk of life think nothing of gabbing on cell phones to friends, family, and business colleagues, no matter whether those others are a mile, a time zone, or an ocean away. Cell phones have become so inexpensive that they're offered for free, complete with built-in cameras, music players, Web browsers, instant messaging, and color video games. Some phones are even disposable.

Such is the revolution wrought by a combination of ultraminiaturized electronics and rapid advances in wireless communications. For good or for bad, the human intimacy of talking by telephone is now within reach of billions of people at all times around the globe.

Yet, as remarkable as the cell phone may be, it is merely the opening act of a much larger and more profound phenomenon. Thanks to the continuing miniaturization of electronic componentry, the airwaves are starting to carry even more chatter, generated this time not by people but rather by potentially trillions of inanimate objects. New digital and wireless technologies are making it possible for virtually every physical object in the world to communicate about itself to interested parties: to other objects, to computers near and far, even directly to business managers and other persons. Objects as diverse as automobiles, jet turbines, candy machines, coffee makers, bus shelters, and cans of pea soup are being assigned unique digital ID numbers, and, over time, many will acquire enough on-board smarts to attain at least bare-bones citizenship on the global Internet. What's more, objects will increasingly "know" just where they are in the world at any given moment, and they will be able to report their locations to remote computers.

Already, consumer products of all kinds may be tracked wirelessly as they leave factories and warehouses, get shipped to stores, or even as they're plucked from shelves and put into customers' shopping carts.

Refrigerated delivery trucks can automatically report to headquarters not only their arrival at a customer's premises but also the temperature of their freezer compartments, the number of items they're carrying, and even how much rubber is left on their tires.

"Smart" vending machines can call for restocking when their inventories run low. Tickets for popular sports events can be protected against counterfeiting by embedding each one with a tiny radio ID "tag."

Racing cars, equipped with dozens of sensors, can transmit reams of real-time data to engineers in the pit stop, even as they're screaming around a track at 180 miles per hour. The list goes on and on.

Indeed, as objects of every kind gain the ability to automatically report their locations and movements and their physical and logical status to computers almost anywhere on earth, entirely new ways of doing business are being made possible. The gap between the virtual world in computer systems and the real world will shrink and then open the door to increased efficiency, new ways of working, and new lines of business. Man's relationship to his tools and to all the things that he makes with those tools is being profoundly altered. Established companies are being enabled to rethink key business processes and squeeze more waste and inefficiency from every aspect of their operations. Equally important, the new breed of location-based technologies is spawning entirely new lines of business and dramatically new business models that were unthinkable just a short time ago.

Consider, for example, what the start-up company Aqui Systems is doing to help drivers stuck in traffic: Equipped with on-board global positioning system (GPS) devices, thousands of vehicles in a metro area can be set up to transmit their locations every minute or two, via wireless links, to the Aqui central computer. There, incoming data is analyzed to determine in near real-time the current flow of traffic on major arteries and to identify potential traffic jams, obstructions, and other problems in their earliest stages. Drivers subscribing to the Aqui service may then proactively be sent wireless alerts and even suggestions for alternative routes—a great help to operators of delivery and maintenance trucks, for instance.

Ten years ago, the Internet and World Wide Web made it possible for practically every person to communicate with any computer and with every other person, thereby unleashing myriad business innovations. Today, the technologies of Real World Awareness are enabling practically every *thing* in the world to communicate with computers, and, as a result, another giant wave of innovation is on its way.

This chapter dissects the innards of Real World Awareness technologies, by using numerous examples of how businesses sense the real world and connect components through networks, all while leveraging a variety of standards.

THE METRO GROUP FUTURE STORE

For now, that innovation is probably furthest along in the retailing industry, even if it's not immediately evident to a casual observer. Wander inside the Extra supermarket in Rheinberg, Germany, for the first time and you would be forgiven for not noticing anything particularly special going on. As in any modern supermarket, row upon row of shelves is brimming with foodstuffs and other goods, cardboard and electric signs promote this brand and that, consumers roam the aisles with their shopping carts, friendly music spills from speakers in the ceiling, and you see a row of checkout counters up front—nothing unusual. Only the seemingly endless variety of sausages and meats at the butcher counter might surprise visitors from other lands.

Look a little closer, though, and you will catch a good glimpse of where technology will be taking retailing over the next decade or so. This store, one of some 2,370 operated by METRO Group (Germany's largest retailer and number five worldwide), has come to be well known in mass-retailing circles as the home of the groundbreaking Future Store Initiative. Behind the scenes—and, in fact, right under shoppers' noses—this seemingly ordinary store has been serving since early 2003 as a living laboratory for testing and perfecting all kinds of Real World Awareness technologies.

Working with 40 technology, brand, and service partners, METRO Group has set out to see how everything from wireless networking to Radio Frequency IDentification (RFID) to mobile computers to the World Wide Web might possibly help large retailers. METRO Group and others of its ilk face enormous challenges these days: withered profit margins, fast-shifting consumer tastes, stiff competition from online merchants, and even stiffer competition from each other. So, as they have often done in the past, big retailers are turning to technology for an edge.

Were a METRO Group executive to offer a private tour of the Rheinberg Future Store, you would quickly become aware of how very "alive" it is with information technologies devoted primarily to the enhancement of Real World Awareness. In the back room, for instance, Extra employees can use a new Web-based kiosk to browse a range of personalized, role-specific information: internal company data, work schedules, policy notices, and even job training courses. Out front, tablet PCs and handheld personal digital assistants (PDAs) connect wirelessly to the METRO Group back-office systems to help staff on the retail floor to answer customers' every question: "When will the new Scorpions music CD go on sale?" "Where's that Jacobs Kroenung coffee I like so much?" "Which Kinderschokolade products come in a family size?"

Indeed, IT is pervasive. Mounted on each shopping cart's push bar is a tablet PC that can often help customers to find what they want on their own (see Figure 3.1). This personal shopping assistant (PSA), as it's called, can display maps of the store's layout to help shoppers navigate their way to any product or category they might seek. By using the store's wireless network, this mobile computer can display pricing information, suggestions for complementary products, and even the shopping lists that consumers may have prepared back home on the METRO Group Web site. A built-in bar code scanner can show the price of any item and keep a running tally as new items get added to the cart. Finally, the screen can display special offers, in-store promotions, and graphically rich advertisements. Because each PSA is electronically aware of its exact location at any moment, these messages can be selected to highlight items directly within a shopper's reach.

Likewise, 19-inch flat-panel screens situated on the walls and shelves of the store can provide helpful information and promotional messages, all instantly downloadable via wireless link (see Figure 3.2). In the drug, detergent, and snack areas, dazzling 42-inch plasma screens show promotional videos and product demonstrations. Customer information kiosks are located throughout the store, too, to provide recipes, product reviews, and even trailers promoting DVD movies.

When it comes time to buy produce, shoppers have only to put their fruits and vegetables on an electronic scale. The device automatically distinguishes between carrots and apples, calculates the price, and prints an appropriate price tag. No bar codes are used; instead, the scale identifies each item by electronically sensing its color, size, and texture.

At checkout time, customers have two options: Use a self-service lane, which requires them to scan each item's bar code. Or, if they have prescanned their purchases in the cart, they can push a button and the cart's total gets transferred wirelessly to a payment terminal.

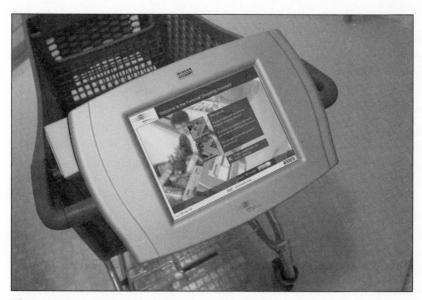

Figure 3.1. METRO Shopping Cart with Display Screen

Figure 3.2. METRO Flat-Panel In-Store Display

Real world and real-time—those are the key concepts, the Big Ideas, in this supermarket of the future. Perhaps nowhere are those concepts being put to a more rigorous test than in the new approaches that the Future Store consortium is exploring in the area of supply chain management. Here, the crucial technology is RFID, which in 2004, from a strategic point of view, was the single most important Real World Awareness technique. RFID enables the precise, accurate, and real-time tracking of goods as they move from factory loading dock to distribution center to warehouse to retail store to a particular shelf in that store—and eventually, right on through to the shopping cart and checkout counter.

In RFID-equipped companies, workers no longer have to manually scan bar codes or punch data into terminals. Instead, electronic scanners situated in key locations throughout the supply chain can automatically sense the arrival and movement of pallets and individual items and instantly send the data they collect to any computer that needs it. With that kind of real-time tracking, producers, distributors, logistics providers, and retailers gain better information with which to make better decisions:

They can recognize and react more quickly to changes in supply and demand, keep shelves better stocked with the right goods, fine-tune prices and promotional messages, and generally respond more effectively than ever to anything that comes their way.

THE MOVING PARTS OF REAL WORLD AWARENESS TECHNOLOGY

To fully grasp the power and potential of the Real World Awareness idea and how it might be applied in any particular industry, you should have a basic understanding of the underlying technologies for sensing, connecting, and responding. These technologies are simple in concept, even if a great deal of research and development work has gone into making them as commercially attractive and easily used as they are today. What follows in this section is an overview of the most important technical elements of Real World Awareness to sense and connect. (Chapter 4 covers the elements used to respond.) This discussion is accompanied, for the sake of those who are interested in the nitty-gritty details, by some in-depth descriptions.

At its most fundamental level, you can think of Real World Awareness as any technique that enables computers to obtain more or less real-time information about the locations, the movements, and the states of objects out in the real world somewhere and to react to this information appropriately. (In some cases, it is individual persons whose location or state—their bank account balance, for instance—is being tracked.) In most situations, this information is collected by attaching some kind of sensor, or tag, to each object and using wireless scanners, or tag readers, to briefly communicate with that tag as the object gets moved from place to place. Tag readers must be in use, therefore, at each and every location where the object's arrival, presence, or departure must be monitored—at each of the entrances and exit doors of a distributor's warehouse, for instance, or in the hands of workers on a factory floor, or at the loading dock of a retail store. Figure 3.3 shows an overview of how Real World Awareness works from end to end.

Another type of Real World Awareness, in fact, involves objects that are "smart" or active enough to be aware, as it were, of their own locations, movements, and states and to make decisions accordingly. These

objects have either enough on-board computing power to make decisions of their own, or they have a link to some remote computer that
can make decisions on their behalf. A vending machine, for instance,
may have sufficient intelligence to keep track of its inventory of candy
bars or soda cans. Based on this data, the machine may alter the advertising messages it displays on its built-in flat-screen display. It may even
send a request to its owner to hurry and bring more Snickers bars for it
to sell.

No matter how it's accomplished, determining the location of an
item or vehicle can be useful in many ways. The vehicles used by field-
service technicians can use on-board GPS to determine just where they
are. Then, dispatchers with access to that data—beamed to them via
satellite or some other wireless link—can then make sure that any new
repair calls are responded to in the optimal fashion. The technician near-
est to the customer, who is free to make a visit, and who has the right
spare parts and tools on his truck, would be assigned the job. Likewise,
the routes of trucks that deliver and pick up parcels can be optimized
by using real-time data about their locations, whether they're operating
in a single city or across an entire region of the country.

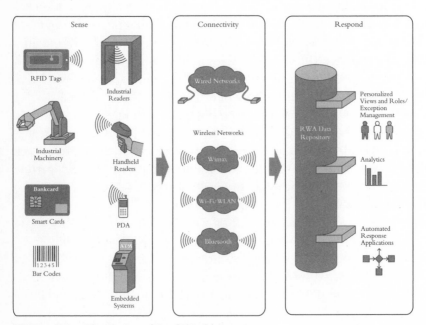

Figure 3.3. The Basics of Real World Awareness

Cellular networks, meanwhile, can now determine the locations of individual callers. Retailers can use that information to send precise, location-based advertising: "Fifty cents off a latte at the Jumpin' Java coffee bar just around the corner." Or, as is being pioneered in Korea and Japan, individuals can pay for cell-based services that alert them whenever anyone on a personal "buddy list" happens to be within a few blocks of them.

There's even a mix of so-called blogging and positioning technologies. In some cities, services are popping up that enable individuals to post what can be thought of as virtual sticky notes, attached to any location in the streets. For instance, someone who has strong feelings about a restaurant they have just eaten at can write a short review on their handheld location-aware device (typically, a wireless PDA) while standing on the sidewalk outside. When the review is submitted to a remote server, location data is automatically attached to it. Later, others who happen to find themselves in the same neighborhood may browse reviews of nearby eateries, perhaps sorting them by cuisine first: "Show all well-reviewed Asian restaurants within three blocks of here." Another proposed use of this technology is to turn the city into a sort of living museum, with everyone invited to post place-specific notes, which might include recorded audio comments, about their experiences and memories.

Radio frequency identification, or RFID, is easily now the most important and commercially promising type of Real World Awareness technology. In its basic form, RFID provides a way to identify individual items and distinguish them from each other and, in turn, to track their location and movement. Advanced forms of RFID, just now starting to hit the market at reasonable prices, are making it possible to determine from afar not only an item's identity and location but also some aspects of its physical state: a shipping container's internal temperature, for instance, or the rotational speed of a truck engine on the highway. The objects being tracked may be complete pallets of consumer goods arriving at a warehouse, expensive medical equipment used in a hospital, trucks moving in and out of a freight yard, or perhaps individual products, like six-packs of beer, bottles of detergent, or packages of razor blades.

Isn't this what bar codes do, you might ask, using just a few stripes of black ink? Why all the fuss over RFID? Indeed, bar codes have long helped make the computer-based tracking of items possible, through every stage of the typical supply chain. And, that technology is hardly about to disappear overnight. Its extremely low cost will be tough to beat in many applications. But RFID offers some significant advantages:

To capture the data encoded on a bar code label requires that the label itself come within a certain distance and in direct line of sight of an optical reader. This ability is necessary because to read a bar code requires scanning its inked pattern with a focused beam of laser light.

As shown in Figure 3.4, RFID tags, on the other hand, need only to be reached by radio waves for their data to be captured. And, radio waves, it turns out, are effective at greater distances than the bar code scanner's laser—at least a few feet and in some cases 10 yards or more. What's more, radio waves can pass unobstructed through many of the common materials used to produce and to package consumer and industrial goods—paper, cardboard, plastic, Styrofoam, and wood, for example. So, in a factory or truck-loading situation, much less effort, if any, is required in order to position each object in a precise way just so that its RFID tag can be read.

RFID: HOW IT WORKS

Attaching RFID tags to pallets of goods and even to individual goods can help both manufacturers and retailers to better monitor and manage the supply chains that tie them together. Rather than rely on workers to manually enter information into terminals, or to scan bar codes, at each step along the way between factory and retail shelf, as is the case now, RFID technology makes it possible for data to be collected automatically (see Figure 3.5). That means fewer data-entry errors and the potential to collect much more data, thereby providing both parties' information systems with a more accurate representation of what's happening anywhere in a supply chain.

Here's how it works: The RFID tag itself is essentially a small, self-contained sliver of electronic circuitry. One element is designed to store some digital ID information, and another element is designed to reliably transmit that ID at the moment a nearby RFID reader asks for it. Radio transmission always requires a certain amount of electrical energy, but rather than use a battery to store that energy within the tag, the tag's circuitry is designed to soak up the energy it needs from the radio waves beamed its way by the RFID reader. Thus, electronically, the tag remains asleep, entirely inert and inactive, until the moment it is bathed in radio waves, at which instant its circuit springs into action and automatically transmits a blip of prestored data to the reader that's pointed its way. For this reason, RFID tags that are powered by external beams of radio waves are called *passive* tags.

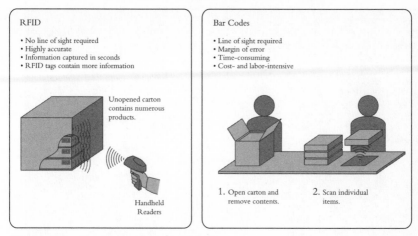

Figure 3.4. Bar Codes versus RFID Tags

Now, imagine a pallet filled with packages of cosmetics, for instance, being wheeled onto a truck as it leaves the factory for a particular retail outlet (see Figure 3.6). As the box is wheeled through the loading dock, an RFID scanner mounted near that door is sending pulses of radio waves several times a second. Suddenly, the radio waves find their target: an RFID tag firmly stuck to the side of that pallet. In a flash, this radio energy triggers the tag's circuits to wake up and respond over the air-waves with the tag's unique ID number. As soon as it receives the ID number, the reader relays the data, accompanied by data indicating the time and place of the scan, through a traditional wire-based network to the maker's enterprise IT systems.

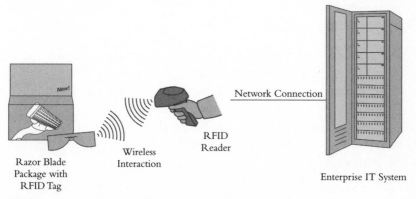

Figure 3.5. Collecting Information from Tagged Razor Blades

Figure 3.6. Pallets Passing through an RFID Reader

Any number of applications may be ready to receive, interpret, act on, and report on this piece of breaking news that's being reported from the field: Pallet No. 314159 passed through loading dock No. 17 at precisely 1532 GMT, Tuesday January 14. Notice of the shipment might be forwarded by Gillette to the retailer's business systems, too, thereby enabling them to alert the appropriate people that the shipment is on its way.

Soon, the box of razor blades arrives by truck at the retailer's distribution center, and, there too, an RFID scanner notes its arrival and forwards the data to the appropriate IT systems.

What Do RFID Tags Contain?

Now, you might well wonder how these two companies' systems can correctly interpret the string of data stored in that tiny RFID tag. After all, the cosmetics maker works with many stores and it cannot afford to create a special set of numbers for each and every one of those retailers. Likewise, each retailer sells goods supplied by thousands of different manufacturers, each of which, not unreasonably, has its own IT setup and unique numbering scheme.

The answer to this conundrum is the electronic product code (EPC), which provides a universal product identification and numbering system that has earned widespread support around the world—hence its common name, EPCglobal. It is indeed open for use by all manufacturers and retailers. It is comparable to, but much more comprehensive than, the ISBNs found on the back of virtually all books these days. Those numbers indicate the country where a book has been published, the publishing company, and the book's category, but all copies of a book have the same ISBN.

Likewise, EPCglobal provides a way to uniquely identify a product (for example, a six-pack of Fizz ginger ale in 12-ounce bottles), the product category of which it is a member (soft drinks), and its manufacturer (Fizz-o-La). But the EPC Global scheme goes a major step beyond that by also providing a means of uniquely numbering individual instances of each product. Thus, each six-pack of that Fizz ginger ale might be assigned its own unique ID number, which would make it possible for any distributor or retailer to keep track of a specific item as it moved through a supply chain.

EPC codes provide a potential way of uniquely identifying all the billions of items that get produced every year in virtually every nation of the world. It is perhaps the most important of many different standards shown in Figure 3.7 that enable Real World Awareness to work. As we explore toward the end of this chapter, the advent of EPC is the key to unleashing the full potential of RFID technology. Suffice it to say that *without* such a universal numbering scheme, RFID would end up being a Tower of Babel, with every enterprise more or less isolated from all others in terms of information sharing.

RFID and Smart Shelves

Eventually, the pallet of cosmetics would get broken apart and individual packages moved to a shelf in a retailer's store. The use of RFID would probably end here, if this were an ordinary store. But, of course, if it were the METRO Future Store, for instance, that shelf might be equipped to work with RFID tags, too. Shelves are being developed that can read the RFID tags on individual items they are holding. The purpose? Whenever an item is removed from the shelf, ostensibly by a customer who intends to pay for it, the retailer's IT systems can be alerted immediately.

If it appears that the shelf is nearing empty, a worker can be dispatched to get the right goods and fill it up again.

This shelf could do more, though. Imagine that a certain item is put on multiple shelves, here and there in a large store, and that at each location a different promotional sign is put on display. Now, because the shelf itself reports on the movement of goods into shopping carts, the retailer's computers would be aware of where in the store each consumer obtained her items. Inferred from that information, perhaps that evening or the following day, may be useful insights into the relative effectiveness of the promotions, the differences in consumer traffic between different locations in the store, and the make-up of shopping baskets—the combinations of items that consumers frequently buy during a single store visit.

Consider this: The smart shelf can easily detect not only when an item is put back on the shelf, but also when it gets removed. If that returning of a particular item is seen to be happening quite frequently, further investigation may be warranted. It may even turn out that many consumers are returning the items within, say, 30 seconds of first taking them into their hands. This may be an indication that the product's packaging is inconsistent, for instance. At first glance, the product may strike consumers as just what they're looking for, but a closer inspection at the description and instructions on the back panel leads them to change their minds and decline the purchase.

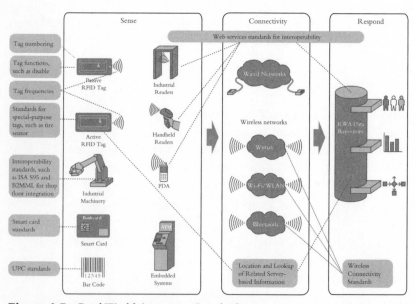

Figure 3.7. Real World Awareness Standards

RFID beyond Retail

Though tiny in size, the RFID tag will have huge impact on all areas of the supply chain, merchandising strategy, and day-to-day operations. And, it will make a big difference in many other industries. Consider these examples:

- More often than you might want to contemplate, surgeons sew up their patients without removing all the instruments and sponges they have used during their operations on those patients. Although detailed protocols have been adopted to fully account for each and every item inserted into and removed from patients, mistakes still get made—and patients still suffer the consequences. Recently, patents have been filed on the idea of tagging instruments and sponges with RFID tags so that those items could be detected inside a patient's body via electronic means. Used as a double-check after a visual inspection and other procedures, it could greatly reduce the risk for all involved.

- Delta Air Lines has begun attaching specialized RFID tags to its aircraft's engines. During each flight, these tags collect data from sensors that monitor the performance of each engine. After the flight, the tag gets scanned and its information sent to a computer for analysis. Using statistical methods that compare the latest data to historical data, engines that need fixing can be identified before they become troublesome. This kind of predictive maintenance has helped Delta to halve its maintenance costs, and it may form the basis of a maintenance service that Delta would offer to other airlines.

- Responding to pressure from retailers such as Wal-Mart and from federal authorities, Purdue Pharma L. P. is one of several companies that are starting to tag the bottles of certain pills it makes. Purdue is the maker of Oxycontin, a painkiller that has been found to be extremely popular on the black market. By law, retailers have to account for every Oxycontin pill they handle, which requires much manual labor and cost. With RFID tags on individual bottles and smart shelves holding those bottles in a pharmacy, it becomes possible to count how many times a day each bottle has been removed from the shelf. If that number doesn't match up to the number of authorized Oxycontin prescriptions the pharmacy has been asked to fill, a problem may well need investigating.

- The shop floor is another place where RFID tags will make a big difference. In most factories, the minute-by-minute coordination of raw materials and parts, work in progress, and work teams gets handled by manual methods: chalkboards, telephone calls, and hand signals. When something goes wrong—a subassembly proves to be faulty or unusable, for instance—precious time is often lost while managers figure out where to reroute work and how to keep production moving forward. By tagging parts and assemblies, however, it becomes possible to keep track of the exact locations of all relevant items and to quickly make decisions about where they should be sent next.

It's nigh impossible to foresee all the applications for RFID that will emerge in coming years. You can be sure, though, that RFID has a role to play in virtually every industry—and, as we will explore in the next section, that much innovation is going on in basic RFID technologies.

RFID TAGS IN DETAIL

Quite clearly, RFID has major implications for the future of virtually every kind of business, from industrial manufacturing to the brewing of beer to the organization and storage of papers in law offices. Recognizing this, the electronics industry has developed a wide range of technologies for use in RFID, each one answering a different need. Some are the size of small coins, and some are even of miniscule size, as shown in Figure 3.8. Radio communication is an inherently tricky technology, though, always forcing engineers to make trade-offs between operating range, power consumption, size, data capacity, and cost, to name just a few of any design's most important parameters. If nothing else, RFID is still a technology in progress, destined to benefit from future improvements in semiconductor, antenna, materials, and fabrication technologies.

What goes into an RFID tag and how it gets used varies widely, depending largely on the type and the amount of data it is intended to provide. The parts of a basic RFID tag are shown in Figure 3.9. In many cases, the information such a tag carries will be identical to that which bar codes based on the universal product code (UPC) standard have provided for the past 30 years; namely, a number perhaps 10 or 12 digits long that identifies the tagged item as a certain type of product—a six-pack of Beck's beer in 12-ounce bottles, for example. Other RFID schemes, however, may identify individual items whose movements must be tracked over a certain period: a specific shipping container or a specific pallet of boxes, for instance. So, although all 12-ounce six-packs of Beck's might get assigned a common ID number, each pallet containing many such six-packs could be tagged with its own, unique number.

Different Tags for Different Purposes

With both six-pack and pallet, however, the tag's ID number serves as a key, or locator number, with which to find additional information about the item—information stored in one or more computer databases, that is. Looking up a pallet's ID number in a business system, for example, might reveal where the beer was brewed, the date of its brewing, its expiration date, and perhaps even records describing each stage of its journey so far: the specific truck that picked it up from the brewery, the warehouse where it was stored for a week, and the cargo ship that hauled it across the ocean. Clearly, there would be no way of recording all that information in a tag whose total memory capacity is just 12 numerical digits. Advances in underlying technology, however, and the creation of tags with their own power and more functions—*active* tags with the parts shown in Figure 3.10—will likely progress rapidly. This section describes the details of these exciting developments.

Reading and Writing to More Memory

Some RFID tags can store only a fixed, preset ID number, and others, costing more and relying on different internal circuitry, permit their internal data to be changed (see the sidebar "RFID Data Storage," later in this chapter). An RFID reading device, in other words, can update RFID data on the fly, just by sending encoded radio waves that cause the tag to rewrite its on-board memory. Indeed, some *rewriteable* tags, as they're called, can accept partial updates; their memories are divided into sections, each of which may be updated independently of the others. One potential application: Members of a supply chain can read the data that upstream partners have written on an item's tag, but cannot change it. Yet, they can also write to only a predetermined portion of the tag for the sake of downstream partners.

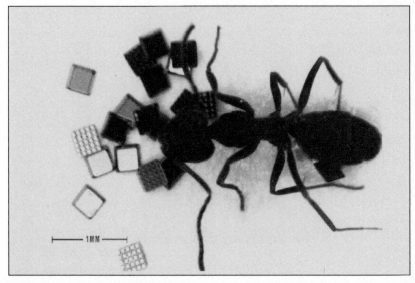

Figure 3.8. Miniature RFID Tags

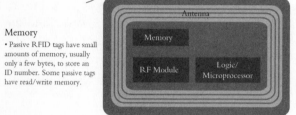

Power
• Passive tags are powered by the energy sent from radio waves to the tag from the reader.

Antenna
• When the antenna receives radio waves in the right frequency, the tag uses the energy to wake up and respond by sending information to the reader.

Memory
• Passive RFID tags have small amounts of memory, usually only a few bytes, to store an ID number. Some passive tags have read/write memory.

Logic/Microprocessor
• The logic on the tag responds to instructions sent to the reader about what information to send back or how to manage collisions.

RF Module
• The Radio Frequency Module makes sense of the signal sent through the antenna and uses the antenna to send information back to the reader.

Figure 3.9. A Basic RFID Tag

In fact, many RFID tags have sufficient capacity to store not simply an ID number but also much additional data about the items to which they're attached. Such a tag might record the date and location of an auto engine's original manufacture and the customer and car type for which it's destined. A beverage keg's RFID tag might record the kind of drink—beer, milk, or soda—that it was used for most recently. This data might help determine, based on local health regulations, the types of beverage that that keg would be legally permitted to carry next.

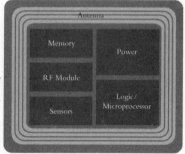

Antenna
• Antennas on active tags may be able to send and receive from greater ranges on many different frequencies.

Memory
• Active RFID tags may have substantial amounts of memory to record data from sensors or data transmitted to the tag about the history of the tagged item.

Power
• Active RFID tags may have their own power source, a battery attached to the device, or an external power source.

RF Module
• The Radio Frequency Module of active tags may be able to receive and transmit on several frequencies.

Sensors
• Sensors enable active tags to gather more information about such quantities as pressure, temperature and vibration that may be related to the tagged item and its environment.

Logic/Microprocessor
• Processing capability of active tags allows filtering of information collected by sensors, advanced collision-management mechanisms, and a set of complex commands. This level of processing power allows the tag to act as an intelligent device and report only meaningful events. Some active tags have RFID readers in them.

Figure 3.10. Smarter RFID Tags

Nokia Corporation is a world leader in mobile communications, by driving the growth and sustainability of the broader mobility industry. In 2003, Nokia's net sales totaled EUR 29.5 billion. The company has 16 manufacturing facilities in 9 countries and Research & Development centers in 11 countries. At the end of 2003, Nokia employed approximately 51,000 people.

President

Nokia Corporation

Pekka Ala-Pietilä

As President of Nokia Corporation, Pekka Ala-Pietilä has a broad range of responsibilities that cover sales and marketing as well as manufacturing, logistics, and sourcing. Pekka, born in Finland in 1957, joined Nokia in 1984 and held a variety of positions: In 1990, he was promoted to Vice President and took the helm as President of Nokia Mobile Phones 2 years later. In 1999, he was named as President of Nokia Corporation. He has been a member of the Group Executive Board of Nokia since 1992. He holds a master's degree in economics, and an honorary doctorate both in technology and in economics and business administration.

Q&A with Pekka Ala-Pietilä, president, Nokia Corporation

Q: Are any broad themes emerging in the sort of applications made possible by RFID technology and other techniques for Real World Awareness?

A: RFID is generating significant interest in logistics solutions, where it is being used to keep track of shipments throughout the supply chain. Mobile RFID and contactless solutions is another emerging application, which focuses on bringing the benefits of the technology to the end users, both for enterprise users and for consumers. These mobile solutions are as follows:

For enterprises, the main focus will be on the mobile workforce, especially in service and repair, and on the use of RFID readers. Examples of applications are shown in this list:

- Using tags to connect your mobile phone to a Web site that posts the latest service or repair description for the specific work you need to do and sends information to you about previous work done on this equipment.
- Using tags for meter reading. The tag identifies the meter and records the reading to the phone or mobile device.
- Using tags to report time and attendance at remote locations: for example, for cleaning, guarding, and repair, and to enable instant invoicing.

For consumers, the main focus is on usability and convenience. A few examples of key applications are creating a shortcut to services (reading information from a tag on an outdoor advertising poster to download a ring tone or movie clip); transferring items between phones (using a secure Bluetooth connection to transfer a newly taken picture); and payment and ticketing (using a phone to gain access at a subway gate).

Q: What benefits will consumers get from these applications?

A: Consumers get improved usability of phone features and easier access to services and content. When we talk about life going mobile, it is all about people and how they use mobile services to enhance their lives. 3G (3rd generation of cell phone technology) is a key enabler for advanced services, and consumers will see it in new functionality in their terminals, in new and exciting services offered by operators, and in the improved user experience of services. For many, a mobile phone will be the first and only kind of communications device.

Q: What will be the biggest mistakes made in applying RFID for consumer needs?

A: Assuming that all RFID technologies are equal and that they can be used only for tracking.

Q: Given that the phone has become a ubiquitous handheld computing device, what role will it play in Real World Awareness?

A: It is likely that several different RFID technologies will be in use, depending on requirements for each application.

Q: How will privacy and security be protected?

A: Privacy and security are of the highest importance in any consumer-facing applications. Near Field Communications (NFC) as a technology provides the means to protect the privacy of the consumer. The mobile phones will not contain RFID tags, but rather will emulate them electronically based on user settings, and therefore allow the user to decide what, if any, information is shown. Finally, Nokia's mobile phone based on NFC implementation will not support reading distances longer than a few centimeters, which addresses many of the distance-based reading concerns that have been raised.

Q: What are the cost barriers?

A: As with any technology, costs will largely depend on volumes. Initially, with low volumes, the cost is considerable, whereas in high volumes it becomes more affordable. RFID as a technology is not very new and not extremely complicated either.

Q: What, besides costs, are the largest barriers to consumer applications of RFID and Real World Awareness technology?

A: Many of the applications considered require a substantial penetration of RFID in phones. There needs to be quite a few people with RFID-enabled phones before it makes sense for a carrier or advertiser or anybody to put out tags that consumers would read with their phones.

Some tags can store pages and pages of information—tens of thousands of bytes of data, that is. This capability is finding use in repair and maintenance. Rather than require field technicians to keep dozens of maintenance manuals on hand, just so that they're sure to have the right one for any machine they may be called to fix, each machine can store its own manual in a "fat" RFID tag and carry a record of earlier repairs as well. To view the manual, which might even contain diagrams, workers need only bring an RFID-equipped PDA within range and call up the tag's contents to their screen.

Readers for Consumers

Nokia has something similar in mind as it starts to build RFID readers into its cell phones. By attaching information-rich RFID tags to bus shelters, monuments, and other objects, out-of-town visitors could summon onto their phones everything from bus schedules and route maps to brief historical anecdotes, perhaps storing information on the tag or using a list of URLs provided by the tag. In theory, the world might become populated with such tags, providing who knows what kind of information to those with the right equipment in hand. Advertising posters, for instance, could provide cell phone users with electronic discount coupons. Restaurants could put tags near their entrances to provide potential customers with menus and critics' reviews.

More Processing Power

Yet another improvement for RFID tags goes even further by not only storing strings of data but also being able to make logical decisions—to act as a tiny computer, that is. In some tags, this logic is effectively "burned" into the circuitry in a permanent, unchangeable form. In other tags, a simple microprocessor is present, capable of running small programs that can be downloaded to the tag's internal memory. The fixed-logic tags, sometimes referred to as employing a "state machine," have the relative advantages of simplicity, lower cost, and lower power requirements. The tags with advanced microprocessors excel in their ability to accept updates, which can give those tags a longer useful life even as conditions and requirements change. The microprocessor tags consume a good deal more power than their state-machine cousins, however, which has implications for how they get used and packaged.

Sensors

In some cases, these "smart" RFID tags can be set up to collect data from sensors such as thermometers, accelerometers, or pressure gauges. And, as RFID technology advances, it's becoming possible to build sensors directly into tags, thus lowering costs and eliminating the need for technicians to wire separate circuits together.

Power on the Device

One of the most important aspects of RFID technology is power consumption. One major category of tags is active, which means that the tags have a battery on-board to provide their circuits with electricity. On-board power gives these tags the ability to transmit information across wider distances than passive tags, which get their power from incoming radio waves. Active tags can also contain considerably more complex circuitry, including more logic and more sophisticated sensors. Some active tags can operate for years without a recharge, by broadcasting a short blip of data every five minutes, for example.

Sooner or later, of course, every battery wears out and needs replacing or recharging, and the cost of batteries is not trivial. Indeed, when thousands of tags are to be deployed, battery costs alone may be prohibitive. So, a great amount of attention has been focused on improving the technology of passive tags. Though simpler and somewhat restricted in terms of operating distance, these tags appear to be sufficient for the great bulk of RFID applications. And, future advances in technology will surely make passive tags increasingly more attractive and useful.

Directional Sensors

In some applications, it's useful to determine in which direction a tagged object was moving when it came within range of a reader. One solution to this problem is to set up two readers near each other and monitor the order in which a particular tag moves through their fields of operation. Another solution is to use a photoelectric detector to detect the motion of the object while at the same time reading its tag to determine its unique ID.

Frequencies

One key parameter affecting the capabilities of any passive RFID tag is the frequency of the radio signals it uses. Radio signals happen to behave

Infineon Technologies AG was founded in April 1999, when the semiconductor operations of Siemens AG were spun off. In March 2000, Infineon went public and is now listed on the Stock Exchanges of Frankfurt and New York. In 2004, it had approximately 35,500 employees worldwide. Revenues in 2004 were EUR 7.2 billion. Infineon offers a broad range of semiconductors and system solutions targeted at selected industries.

Peter Bauer

Executive Vice President, Chief Sales and Marketing Officer

Infineon Technologies AG

Since 1999, Peter Bauer has been Executive Vice President and Chief Sales and Marketing Officer and a Member of the Management Board. After completing his studies in Electrical Engineering in 1986, he worked in several positions and managed divisions around the globe. Born in 1960, Peter is married and has two children.

Q&A with Peter Bauer, Executive Vice President, Infineon Technologies AG

Q: What's needed to spur RFID's further adoption by corporations?

A: Three factors: lowering the cost of tags, improving their functionality, and, perhaps most important, working out new business processes that can take full advantage of the technology.

The cost of tags is steadily decreasing and will continue at about 20 percent a year. Tags costing 40 cents each today, as used in consumer packaged goods, will fall to around 20 cents a tag over the next few years and perhaps to 15 cents over the long term. Antenna design is a particularly fruitful area of research, as it offers much potential for cost reduction. Certain kinds of tags, built from polymer or plastic, will soon be produced in high volumes to sell for just 1 cent apiece, but they are quite limited in function.

Much progress is being made in improving the functionality of RFID tags. One area of intense research is in getting tags to operate well in the presence of containers full of liquids, such as industrial chemicals or beverages. In general, liquids tend to block radio waves. Another area of research is preventing internal data within tags from being tampered with. This is especially important in the pharmaceutical industry, where the Food and Drug Administration (FDA) requires individual bottles of medicine to be tagged with data about their production batch. The tag data can be encrypted, or the tag may use a self-destructing memory circuit. Future tags will have the ability to permit access to different sections of their memory on a selective basis: Perhaps a manufacturer could "touch" only sections A and B while distributors would have access to section C and retailers only to D.

Q: And business processes?

A: The biggest challenge for enterprises is to make sure RFID gets used to make business processes effective. This is why Infineon works closely with software makers such as SAP, which have a deep understanding of business processes and how best to automate them. Of particular concern is that a world full of RFID tags will generate torrents of new data, which, if not handled correctly, may defeat the very IT systems that a company wishes to help with RFID. Infineon sees an opportunity, however, to actually use the decentralized storage of data in RFID tags as a way to reduce complexity. With increasing levels of computational power available inside each tag, it may filter data itself. Thus, a tag designed, say, to measure the temperature within a refrigerated container would send an alert only when that temperature went beyond some preset range—when the container became too hot or too cold, in other words. We reckon that as little as 1 percent of the information stored in a large set of RFID tags needs to reach the overseeing ERP system.

Q: Does Infineon use its own RFID technology?

A: Of course! Our Dresden, Germany, based microchip plant uses RFID to monitor and manage the flow of semiconductor wafers through a lengthy and somewhat tricky sequence of processing steps. This has enabled Infineon's engineers to greatly reduce manual data-entry tasks, reduce the number of errors entering our IT systems, improve quality control, and cut production costs. Finally, using RFID enables us to turn around new microchips faster than we ever could before. Since October we have started to use RFID labels on product shipments between our production site in Regensburg, Germany, and our European distribution center. With this innovative step we are evaluating the impact on our logistic processes, our IT systems as well as the many improvements gained by using an RFID-enabled supply chain management system. In time we will implement this system all over Infineon as well as connect to our customers and distributors.

Q: What's being done about privacy concerns?

A: There is public concern about RFID encroaching on personal privacy. Privacy advocates have feared, for instance, that it will be possible to read tags embedded in people's clothes as a way to track their physical movements. Or, they suggest, it will be possible to read the monetary value stored in a smart banking card even while it is in a person's wallet or purse. A retailer might use this knowledge to treat their seemingly wealthier customers to better service than those whose cards indicated less value, for instance. These concerns are not to be dismissed out of hand, Infineon believes, but they can be addressed. One important first step is to educate the public with the facts about how RFID technology works and about its technical limitations.

quite differently depending on their frequency. (Consider, for example, the long distances that AM radio stations, operating at low frequencies, can cover—as much as 1,000 miles, at night when interference is reduced—compared to the 50-mile range of FM and TV stations, which operate at much higher frequencies.)

Lower-frequency radio waves can generally travel farther and penetrate solid substances more effectively than higher-frequency signals. This ability can be of help in an RFID setup where it may be necessary to read tags no matter whether they are situated behind a box containing some particularly dense material. It can also help extend the range of an RFID system: Low-frequency signals can effectively reach about 1 meter and still deliver enough power to activate a simple passive tag. On the other hand, lower-frequency signals cannot carry as much information per second as those using higher frequencies.

Demand is increasing, though, for "smarter" RFID tags, with onboard logic, and this requires higher frequency radio links. Their logic may be used to perform a calculation or to compare several pieces of incoming data—collected from a thermometer, for instance—and make a decision based on the results. As noted earlier in this chapter, some tags have logic burned into them, and others contain a simple general-purpose microprocessor. In either case, it is the frequency of incoming radio waves, sent by the RFID reader, that will determine just how quickly a tag can execute its logic instructions. Just as a higher-frequency personal computer (as measured in gigahertz, or billions of internal clock cycles per second) can make a video game program run faster than a slower PC, higher-frequency RFID tags can do more internal "work" and tackle trickier decisions each second than can lower-frequency tags. Rather than use an internal clock, however, passive tags will operate at the frequency provided by the radio signals that are sent their way.

Managing Collisions

A higher radio frequency can help solve one particular tricky technical problem in RFID. When more than one tag passes simultaneously within range of an RFID reader of the sort shown in Figure 3.11, it tends to be quite difficult to single out any particular tag and read its content and its contents only. The RFID reader's beam of radio energy may activate many passive tags at once and cause all of them to transmit at once. With their many signals overlapping, it can be almost impossible to successfully read from or write to any single tag.

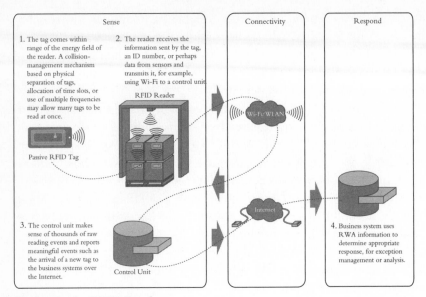

Figure 3.11. RFID Readers

To fix this problem, engineers have worked out a variety of *anticollision* techniques. Tags may operate on different frequencies, for instance. Or, more compelling from the point of view of cost and practicality, the tags may pick up on some split-second timing signals sent to them by the scanner and, using their internal logic, respond in a precisely timed, non-overlapping manner. Clearly, this kind of communications protocol requires some fairly sophisticated and extensive software, and that calls for relatively high-frequency radio signals to speed along the software's execution.

Frequency Ranges

Many early-generation tags have operated at the frequency of 120,000 Hertz—well below the AM radio dial in the United States. But now, one of the most popular frequencies for running RFID tags is 13.56 MHz, or about 13-and-a-half million cycles per second. This frequency is high enough to make a smart tag zip through some fairly involved software routines in a reasonable amount of time. This faster logic, moreover, helps with anticollision protocols and can make it possible to read tags reliably even as they move past a scanning station at higher speeds than was possible before. This ability is particularly useful in logistics, where it's in no one's interest to retard the movement of pallets or forklift trucks.

RFID systems based on *ultra-high-frequency (UHF)* frequencies (868 MHz in Europe and 915 MHz in the United States) offer a longer range and certain other advantages. They can communicate at distances of as much as 3 meters when used with passive tags and as much as 15 meters with active, internally powered tags. That has obvious advantages for certain applications.

A step further up the radio dial is the 2.4GHz band, known as *microwave*, which offers the highest possible data transmission and processor clock rates. Microwave tags can perform quite complex calculations. Chemical containers, for instance, can be programmed to recognize each other and set off a safety alarm if they determine, autonomously, that more than four containers, for example, happen to be located within a certain radius of each other.

This ultrahigh frequency also means that tags and base stations can move quite quickly relative to each other with no loss of reliability or accuracy. Maximum speeds of as much as 400 kilometers per hour (240 mph) have been achieved in highway-traffic-related applications. At this frequency, however, active tags are by and large a must because of high power requirements. The range of these tags is significantly greater, too: 10 to 15 meters is realistic, and even greater ranges have been achieved in the field of traffic management. Automatic toll systems on highways, for example, can use UHF signals to transfer identification data over distances of as much as 50 meters. One negative factor is that UHF signals are even weaker than 13.56MHz signals when it comes to penetrating most materials.

Sensors

Finally, with enough on-board processing power, an RFID tag can serve not only to store and report on internal data but also to collect and transmit "live" information from physical sensors. An entire range of sensors can be built into electronic labels to measure such parameters as temperature, humidity, air pressure, shock, acceleration, and even light intensity and pH values. So far, however, this is more theory than practice, mainly because of issues of cost. Electronic tags with built-in sensors are not now widely used because the sensors themselves tend to consume too much power. Sensors and their associated electronics in a tag can consume anywhere from 10 to 1,000 times more energy than passive RFID tags—those that take their power from incoming radio waves, that is—are able to provide. As a result, sensors have been used mainly in active systems whose tags each have their own on-board batteries.

In short, no matter what the requirements are for a particular application, there's likely to be an RFID tag that's just right. And, if there isn't, a lively RFID industry is emerging that's prepared to design and produce just about any customized tag that might be needed.

RFID DATA STORAGE

Different storage technologies have different properties and, no surprise, different costs. In many RFID applications, simply being able to read the data stored in a tag is sufficient. In some situations, however, it's helpful to be able to remotely write new data to a specific tag. As a finished product leaves a factory, for instance, the date of its production may be written to the tag.

Not surprisingly, a storage medium that can accept updates will be somewhat more complex and costly than one that permits only reading. To write data into a semiconductor memory, for instance, requires a certain amount of electrical energy, so writeable tags, as they're sometimes called, must either contain a battery to supply power or use circuitry that can extract the required energy from incoming radio waves. Batteries, of course, do not last forever, so battery-powered tags have a finite lifetime—sometimes stretching to several years, given the right battery and an application that doesn't require the tag to be activated too often.

There are several varieties of read-only semiconductor memory technologies, each of which is finding use in some corner of the RFID market:

EEPROM: Electrically erasable programmable read-only memory does not require a continuous power supply to store data. The power required to read information from an EEPROM is quite small. Rewriting the data within individual EEPROM cells requires significant energy, however, which limits how far a passive EEPROM-based tag may be from a transmitter. Also, EEPROMs can't be updated limitlessly. Depending on their design, they can accept a maximum of 100,000 to 1,000,000 rewrites.

SRAM: Static random access memory has the advantage of being able to change its contents much faster than EEPROMs. The downside is that they require continuous power, without which they "forget" any data they may have been storing. That means incorporating a battery in SRAM-based tags, which leads to a higher cost and weather-related limits; batteries don't work well in extreme cold, for instance.

FRAM: Ferromagnetic random access memory is a comparatively new technology and is little used. It combines some of the advantages of SRAM and EEPROM.

Memory Capacity

RFID tags vary in memory capacity from a mere 32 bits (equal to 4 bytes, or 4 characters of information) to as much as 32,000 bytes (or 32 kilobytes). In general, the less data a tag carries, the further away from a reader it can be and still be interrogated reliably.

Low-end tags range in capacity from just 32 to 80 bits. These low-capacity tags are designed to store only an identification number, but by using the full 80 bits of binary coding, that number can range from 0 to 1.2 x 1024—enough to give a vast number of items unique IDs.

Another variety of tag stores 80 to 256 bits, using primarily EEPROM and RAM memory technology. These tags can include small amounts of explicit information encoded in a sort of numerical shorthand: Just as airlines refer to the Newark airport as EWR, for instance, these tags might use a few binary 1s and 0s to signify a lengthy product number or customer name. Only a computer with the right decoding software can make sense of such data.

SRAM-based tags offer 32 to 256 bytes of capacity, and FRAM technology can store as much as 32,000 bytes. These types of memory can store information as alphabetic letters as well as binary-based codes. In fact, 32 kilobytes is enough to store some 11 pages of standard, double-spaced text. Such capacity makes it possible to store highly detailed information in explicit form, such as complete maintenance instructions for a component of a larger assembly. Just like the other kinds of memory technology, however, SRAM and FRAM don't escape the challenging technical trade-offs between memory capacity, operating range, and power consumption.

State machine versus Microprocessor

Every RFID tag contains logic circuitry whose job, at minimum, is to control the movement of data within the tag. In sophisticated tags, however, such logic can be used for sophisticated purposes: A tag may section off its total memory and permit only readers that supply the proper pass-words to update particular sections—a capability that might prove useful in a supply chain involving multiple parties. Whether it's simple or com-plex, a tag's logic may be implemented in the form of a state machine or as a programmable microprocessor. Each has its pros and cons.

The state machine is the most popular form of logic in today's tags. State machine is simply a technical term for a circuit designed to per-form some preset sequence of logical instructions—an algorithm, that is—and nothing else. The state machine uses little power and is quite efficient in operation, and in high volumes it can be produced at a rel-atively low cost per unit of logic. Its computing power makes it useful for handling such tasks as encrypting data, controlling access to a tag's memory via passwords, and executing the complex handshaking proto-cols that are required in order to prevent multiple tags from interfering with each other when they're all scanned at once. The state machine has the distinct disadvantage, however, of being hard wired; its logic cannot be changed after fabrication.

Enabling that kind of logic modification is exactly where the micro-processor shines. It is a general-purpose computer, in essence, that can execute any string of software instructions it is given. With a micro-processor, an RFID tag can perform fairly sophisticated tasks, such as monitoring and analyzing inputs from several outboard physical sensors. In some applications, such a tag may actually run its own tiny operating system, which could be useful in making the tag's software, as well as its data, modifiable from afar. **Tags used for tracking maintenance of equipment in hard-to-reach places is an excellent application of this technique. Readers can send instructions to update tags on such equipment without having to get close to them.** Micro-processor-based tags consume more energy than their state machine counterparts, however, and that can limit their ability to communicate with scanners across long distances.

EARLIER REAL WORLD AWARENESS TECHNOLOGIES

As seen in the METRO Group Future Store, RFID is hardly the only technology contributing to Real World Awareness. RFID probably has the most potential to change how business gets done in the future, but bar codes, smart cards, smart labels, physical sensors, and a variety of wireless positioning technologies are making strong contributions, too. This section looks at each of them.

Intel Corporation, which introduced the world's first microprocessor in 1971, supplies the computing and communications industries with chips, boards, systems, and software building blocks that are the "ingredients" of computers, servers, and networking and communications products. Intel, founded in 1968, had revenues of just over $30 billion in 2003 and employs more than 78,000 people in 48 countries.

Patrick P. Gelsinger

Senior Vice President and Chief Technology Officer

Intel Corporation

Pat Gelsinger is Senior Vice President and Chief Technology Officer of Intel Corporation. Pat joined Intel in 1979 and has more than 20 years of experience in general management and product development positions. Pat leads Intel's Corporate Technology Group, which encompasses many Intel research activities, including leading Intel Labs and Intel Research and driving industry alignment with these technologies and initiatives. As CTO, he coordinates Intel's longer-term research efforts and helps ensure consistency from Intel's emerging computing, networking, and communications products and technologies. Gelsinger has electrical engineering degrees from Lincoln Technical Institute, Santa Clara University, and Stanford University.

Q&A with Patrick P. Gelsinger, Senior Vice President and Chief Technology Officer, Intel Corporation

Q: What sort of RFID-related products is Intel building?

A: RFID is a range of devices, from the most simple passive tags that just report an ID number to these active sensors that have quite a bit of computer and communications power. Right now, Intel is not building the RFID tag component itself, but we are building every other element involved in making RFID work. We're building RFID readers, and we are also very involved in research around future usage models for RFID in the medical area, for medicine bottles or home care.

When most people think about RFID today, they think about passive tags. We are very, very involved in active tags, which we call active sensors. They have a small battery, a communication processor, and a sensing device—for example, an accelerometer for motion, or a thermometer for temperature, etc.

Q: What different roles will passive and active tags play? How will they work together?

A: One issue is cost. There are going to be things that the passive tags do just because of price that the active tags never will. If the passive tags are at 20 cents or so now, you know, the cost of an active sensor is probably 10 dollars, but with volume it should get to be a couple of dollars.

You may take one of these active tags, put a larger battery on it so that it lasts 10 or 20 years, and add a very sophisticated accelerometer or other environmental sensing capability. A $100 sensor with a $2 or $3 communication computer might be atttached to a million-dollar piece of factory equipment. The cost of a sensor is nothing when I consider that I'm now trying to sense and understand maintenance associated with the million-dollar piece of equipment.

The amount and quality of data that's being delivered will be tightly linked to the value of the item being tracked or monitored. For a retail item, the tag will probably have to be 10 percent of the cost of the retail item, preferably on the order of 1 percent. But an active device will monitor larger units of value. I may put it in the pallet full of retail items that each has its own, passive tag. The pallet of them itself may be worth a hundred or a thousand times the price of any individual item.

An active tag on a pallet might also have a reader to sense the passive tags of the items on the pallet. By the time I actually get to the warehouses, the active tag is going to tell me not only that I have 1,246 things on this pallet, but also the temperature, how long it has been in transit, and how much movement it has had in transit. If, in fact, the pallet might have been programmed with the environmental range that guarantees a safe delivery by the shipper, it might include GPS as well, and I'll know all the places that it went to along the route. When the pallet arrives at its destination, the systems will have to read only the active tag once instead of the tricky problem of reading all 1,246 individual tags. And because the tag is active, it will be able to be read from much greater distances with wireless networks.

Q: What role will wireless networks, like WiFi and WIMAX, play in the expansion of RFID?

A: To us, the first step of this is just getting the data. The second is to accumulate that data. We're building access points today that have the computing power to gather, store, and process information. Then the question is "How do you get it back to the Internet?" Wireless networks will be used differently based on distances. WiFi hot spots in factories might be the way to collect information from readers, but then the WIMAX—which can travel much larger distances, up to 50 kilometers—might be used to get the information to a collection point connected to the Internet. We don't see WiFi competing with WiMAX per se. We see it very much as local connectivity versus metropolitan or regional area connectivity.

Q: How will the adoption of RFID progress?

A: RFID is going through the typical hype curve, where at first people claim it is going to transform the world overnight. Then, when people try to implement it, they come back, painfully disappointed. Then, over time, it reaches a maturity and acceptance level that will be dramatic and significant beyond even the expectations of the initial hype, but often in places and ways that weren't quite envisioned or considered initially. Finally, there is just this maturity-of-adoption cycle that is going to have a huge impact. At this point, I think we have sort of gone past the peak of the hype.

So, our long-term view is a combination of RFID as well as the active sensors that we have been discussing today, working in combination and conjunction with each other, being deployed in many billions of items as we go into the next decade. A lot of those struggles to make everything work in a meaningful way have nothing to do with the underlying RFID technology itself, but all of the integration process, refinements, and changes are required in order to gain full benefit from the technology.

Bar codes, in use since the early 1970s, based on the UPC and other standards, now appear on virtually every retail item in stores, from newspapers to cereal boxes, beer cans, and shampoo bottles. The technology has found widespread use in industry, too, helping with all modes of logistics. Bar codes typically encode numerical data as a sequence of inked bars and blank spaces that vary in width. Some bar codes are of fixed length, and others vary depending on how much data they need to convey. Certain two-dimensional bar codes, using a checkerboard pattern of square dots in a rectangular field, can store more information. They have found use by the U.S. Postal Service and other shipping companies, among others. Bar codes can be printed either directly on a product package or on a paper label with adhesive backing.

The bar code reader scans the label's parallel bars with an oscillating laser beam. A photodetector records the reflected light, and a microprocessor decodes the pattern of varying brightness into a series of binary 1s and 0s. A variety of bar code standards define specific rules for coding data this way, each one aimed at a particular industry or vertical market. On the European consumer goods market, for example, EAN code 128 is the standard of choice.

Chip cards, or *smart cards,* are electronic memory devices packaged in the form of a credit card-size piece of plastic. They were first used widely in the mid-1980s, mainly to create prepaid telephone calling cards. Since then, the technology has found use in the banking, airline, and security industries (see Figure 3.12). The cards store information in a standard flash-memory chip—able to store information without continuous power—that is mounted just below the surface of the plastic. On the surface, an array of paper-thin metal contacts can connect the chip to a set of electrical probes within an appropriate reading device. Data can then be read from the memory chip and, in some cases, written to it, too.

A second generation of chip cards, referred to as smart cards, adds a built-in microprocessor for on-board computing. Some health insurance companies issue simple smart cards to identify clients when they visit a doctor or enter a hospital. By 2006, German citizens will carry health cards that store, in heavily encrypted form, data about their past illnesses, allergies, and other medical facts. Smart cards are finding use as electronic wallets, too, to store a cash value that consumers can spend at local merchants, and as telephone cards within GSM-based cellular phones. In many smart cards, the microprocessor handles encryption routines designed to help keep the card's stored data secure against unauthorized access or tampering.

Figure 3.12. A Smart ID Card

Smart labels, a particularly cost-effective tag for high-volume applications, appeared on the market a few years ago. The smart label is similar in shape and size to the smart card, but it is much thinner and more flexible. A flat, paper-thin coil antenna and transponder IC (integrated circuit) are connected to each other and usually affixed to a piece of flexible material. This device can then be laminated to an adhesive-backed paper label and affixed to an object or even sewn directly into a garment.

A variety of other *miscellaneous Real World Awareness electronic technologies* can locate items in physical space. The retail industry, for example, has long used electronic article surveillance (EAS) tags to protect against theft. A tag firmly attached to a pair of jeans sets off a loud alarm if the tag has not been deactivated and passes through a detector located at a store's exit. Ultrathin antitheft tags have been developed for pasting inside books and audio CD packages, too.

The need to track high-value assets has driven the development of even more location schemes. A setup named Lo-Jack, for instance, relies on a radio transmitter to help locate stolen cars. Satellite setups help freight companies locate and communicate with 18-wheel trucks out on the highway.

Cellular-phone-based transponders are used to track shipping containers and vehicles. The transponders rely on the same GSM transceivers found inside cell phones and deliver information in standard cellular form. The big advantage of this kind of location reporting is that it works wherever a cellular signal is available, which is almost everywhere. After the information is in that network, it can be delivered to any spot where the telephone network reaches. The downside is that each communications session incurs the cost of a regular telephone call.

Cellular companies are adding facilities to their base stations, too, that enable them to locate individual mobile phones while the phones are active. Depending on the terrain involved (urban canyons, where radio signals tend to echo wildly, are more challenging than wide-open landscapes), accuracies of as little as 10 meters are possible. The positioning is accomplished by measuring the timing of a phone's signals as they reach different base stations. One of the first uses of this kind of positioning has been to help police and ambulance services quickly locate individuals who have called 911. But the technology is starting to spawn a variety of advertising, dating, and other types of services.

Other positioning technologies are afoot. One of the latest methods for locating items in and around office buildings and public spaces is to use a tag that relies on the signals already radiating from standard wireless local-area networks (LANs)—often called Wi-Fi. Here, too, costs can be relatively high because the required tags are fairly sophisticated and not yet being produced in much volume. In metro areas, meanwhile, determining locations can be done through the use of standard broadcast television signals. One big advantage of this approach is that such signals are already present in most major markets and they reach locations, such as inside office buildings, where traditional GPS technologies fail. By adding precise timing data to existing TV signals, what's more, accuracies can be greatly improved. A company named Rosum has recently raised venture funding to commercialize this technology.

NETWORK TECHNOLOGIES

As you have seen, clearly, one of the most critical elements for giving enterprise information systems Real World Awareness is the ability to send and receive information to any spot where it's being collected or needed, respectively. (Figure 3.13 shows the network technologies involved in Real World Awareness.) Fortunately, wireless communications technologies have enjoyed tremendous progress in recent years, which makes it possible to untether most kinds of computer equipment and establish high-bandwidth links between almost any two machines no matter where they might be located. Even computers and sensors that are moving can be reached wirelessly. Without recent advances in wireless connectivity, the revolution in Real World Awareness would be proceeding at a much slower rate, retarded by the difficulty of bridging the gap between sensing events and communicating with business systems that can respond to them with wired networks.

Keep in mind that innovation in wireless technology is moving at a terrific pace right now, which makes it quite difficult to predict just how specific wireless and wire-based communications technologies and services may compare with each other in the future. As computing power gets cheaper, as new techniques are worked out for encoding radio signals with digital information, and as antenna technology improves, it's becoming possible to cram more and more information onto the airwaves at ever-diminishing cost. Meanwhile, new ways of employing radio links are making possible powerful new networking methods. One of the most promising is mesh radio, which aims to re-create the highly decentralized—and therefore highly reliable—Internet in the air, with multiple pathways connecting every node to every other node.

It can be useful to think about the various wireless networking schemes that are available as a set of concentric circles, as shown in Figure 3.14. Each circle represents a class of network that is distinguished from the others mainly in terms of its maximum radius of coverage. Because of the nature of radio propagation, it's not economically feasible to use a single wireless networking technology to span both distances of only a few feet and distances of a few miles. Instead, different technologies, which vary by radio frequency, power output, and other technical parameters, are being developed to cover each range. Naturally, some overlap occurs between these different schemes, particularly as their underlying technologies and components advance in function and cost, but they do fit into a fairly simple model.

The smallest, center circle represents the *personal area network (PAN)*, whose job is to connect devices that people carry on their persons or use on their desks, for instance. In short, PAN networks strive to replace the jumble of cables that would otherwise be necessary to connect these devices. Such a network might connect a PDA to a cell phone, for instance, thereby enabling the PDA to link through the cellular network to the Internet. A PAN could connect a keyboard to a desktop computer, or a laptop to a nearby printer. Some personal headsets even use PAN links to connect to the cell phones hanging on their users' belts. In general, PAN setups don't operate at the highest connection speeds, but they excel in using minimal power and enabling many connections to operate without interfering with each other.

Real World Awareness Connectivity

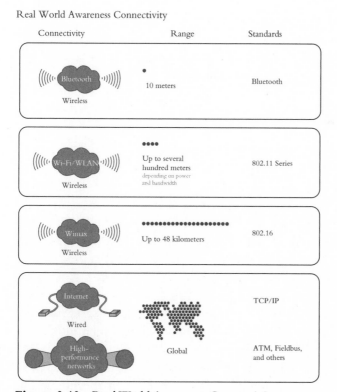

Figure 3.13. Real World Awareness Connectivity

Global Wireless Standards

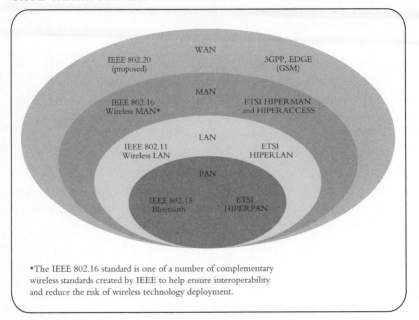

Figure 3.14. The Scope of Network Technologies

BLUETOOTH

Evidently, the most mature and widely accepted PAN technology right now is the standards-based scheme named Bluetooth. It enjoys the backing of many large companies, which have brought out a wide variety of Bluetooth-compatible products. One of the technology's most attractive features is its low consumption of electrical power, which makes it well suited for use in portable, battery-powered devices. Bluetooth standards have also been worked out for around 29 classes of applications, such as controlling a fax machine or operating a hands-free headset. These standards make it easier for manufacturers to bring new products to market and ensure that they will work with Bluetooth gear from other makers.

A number of other PAN technologies are afoot, which may eventually compete with Bluetooth in certain applications. One is a scheme named Zigbee, aimed primarily at low data-rate applications, such as sending control signals in industrial plants or even in the home. For instance, a central control computer might communicate with a family's refrigerator, air conditioners, and heating system via Zigbee, thereby enabling less consumption of energy. Another wireless technology, able to cover short distances at extremely high data rates, if needed, is *ultra-wideband (UWB)*. In theory, UWB will be able to send data at rates of as much as 100 megabits per second over distances of up to 50 feet or more. That's enough to feed high-definition television data from a TV receiver to a flat-screen display on a wall at the other end of a room, for example. Like Bluetooth, UWB offers a way of reducing the rat's nest of cables that typically show up behind computers and home entertainment systems.

Wi-Fi/WLAN

The next-largest circle in our conceptual diagram is the *wireless local-area network,* called both *Wi-Fi* and *WLAN,* which now is the most widely deployed type of wireless data network. The wireless LAN typically acts as a wireless version of the popular Ethernet local-area net, which connects laptops, desktops, and other devices to each other and to a node on the Internet within the confines of an office or home. Recently, WLAN access points have started to appear in public "hot spots," both indoors and outdoors, to give mobile users access to the Internet from their laptops and PDAs. WLAN technology is also coming to be used as a medium for carrying telephone calls, video streams, and other types of information. It is likely in the near future that just about every indoor location where an RFID reader might get used, whether it's handheld or fixed to a wall, will be equipped with a wireless LAN. Proposed extensions to the current WLAN standard will increase the maximum data rate from a current 54 megabits per second to 100 Mbps or more.

WiMax

The *metro-area network (MAN)* generally covers an entire town or city, or some part thereof. The main technology to keep an eye on is *WiMax,* which has been the focus of much development activity. It will permit roaming within a radius of a mile or two—enough, anyway, to enable the deployment of RFID readers and other Real World Awareness devices well beyond today's in-premises limits.

WiMax has many potential uses. It could serve as an alternative to the cellular-based links that some companies are now using to communicate with distant machines such as vending machines, vehicles on the road, appliances, certain industrial machines, and even traffic lights and air-conditioning systems. This kind of communications will be increasingly important to corporations as they seek to enhance the Real World Awareness aspects of their information systems. The cellular network reaches many locations, but each call placed over its circuits costs a fair amount of money and the available bandwidth is not particularly high. This could change as 3G technology is deployed, by offering the high-bandwidth transmission of data, but WiMax is likely to be a serious, if not dominant, competitor. If nothing else, WiMax has the backing of Intel, which is funding start-ups to develop many WiMax components and also scrambling to develop its own. Indeed, Intel has announced plans to build WiMax into its Centrino mobile computing platform within a couple of years from now, which will give laptops and other devices high-bandwidth Internet access beyond the limits of today's Wi-Fi local-area nets.

If all goes well, WiMax will follow the same steep price curve that has stimulated Wi-Fi's use. This has been the result of a rigorous standardization effort, which in Wi-Fi's case has been supported by several hundred companies. Over a period of just three years, competition among those firms has driven down the price for residential Wi-Fi access points, the devices that transmit to and receive from any wireless-equipped computers in the vicinity, from several hundred dollars apiece to just $20 now. This price decline has enabled the technology to show up in millions of homes and public areas—airport and hotel lobbies, city parks, and espresso bars—that are equipped with Wi-Fi access. It has helped a great deal, too, that Intel built Wi-Fi into its laptop computer designs, under the Centrino brand name, and Wi-Fi is now showing up in cell phones, which makes it possible to make low-cost phone calls across the Internet.

With 150 companies now backing the WiMax standard, something similar may well take place around that technology, too. Though the technology has been touted mainly as a method for bringing broadband Internet access to homes and offices, in lieu of digital subscriber line (DSL) and cable modem service, WiMax has the ability to provide lower-bandwidth network access to large geographic areas, too.

The outermost circle represents the wide-area network (WAN). It is designed to cover distances measured in tens, hundreds, or even thousands of miles. This technology has been commercialized only a little, mainly for use in connecting buildings located on opposite sides of a city. Used with specialized high-gain antennas, for instance, WiMax links can reach across distances of 30 miles. And, it is possible to communicate even further—over the horizon, even—by using enhanced versions of the basic technology that underlies WiMax.

Specifically with regard to Real World Awareness, WAN technologies will help to extend the reach of traditional RFID setups and to enable virtually any electronic device to contribute to the Real World Awareness of enterprise IT systems, regardless of its location. With a wireless network operating inside a warehouse or the company parking lot, portable RFID readers will remain connected to the warehouse management system, for instance, even as they are moved around. Many such readers will be handheld, designed for use by workers roaming a warehouse floor or parking lot. Already, major electronics makers are planning to include RFID readers in cell phones and PDAs. Within retail stores, wireless links can collect information from RFID-equipped store shelves, shopping carts, and POS devices, and these links can distribute information to those same devices and to electronic signage located throughout the store.

Mesh Radio

Some quite intriguing variations on these basic wireless schemes are just emerging. One of the most promising is *mesh radio*. Its core concept is the creation of a decentralized, Internet-like linkage between many different wireless nodes. On the traditional Internet, devices, called routers, are in place to move packets of information from any two points. The routers are just computers designed specifically for this job. The information about what route any particular packet should take to get from point A to point B is not stored anywhere centrally. Instead, it is worked out on the fly by each successive router that the packet encounters. The routers, controlled by some specialized software, are continually in touch with their neighbors, to collect information about where congestion may be occurring, where a particular router has failed, and where new outbound paths are becoming available.

Together, this distributed intelligence, as it were, makes the Internet particularly resilient and efficient. The network is continually healing itself, finding new paths around trouble spots, and making sure that all packets eventually find their way to their intended destinations.

Mesh radio schemes attempt to do the same thing, but over the airwaves. Wireless nodes, able to send and receive packets to each other over distances of hundreds of yards or perhaps a mile or two, can be set up on residential rooftops, office buildings, or antenna towers. Each node calls out to find and identify its neighbors, and together they all form a network in the air. Data packets get shunted from node to node, much as on the Internet, and eventually find their way to a node that is connected to the Internet proper, where the packets are whisked off to their destinations.

The great appeal of the mesh radio concept is that it enables the wiring of a neighborhood at a relatively low upfront cost. No digging of trenches or string of wires across poles is required. Low-cost mesh radio nodes can be installed one by one, wherever they're needed, to provide service across any area of land. What's more, as each node gets added to a mesh, the aggregate bandwidth of the entire mesh increases, for each node creates more possible paths for data packets to travel across. All that's required is that the nodes have routing software that makes sure packets don't just circulate aimlessly through the mesh but instead quickly find their way to an existing node and zip off through the Internet.

The mesh concept is just catching on as a potential way to wire up suburban neighborhoods for broadband Internet access, in competition with cable and digital subscriber line (DSL) services. Something quite similar is also at work in the sensor market, under the name Smart Dust, and it has major implications for Real World Awareness. The basis for that name is the fact that it's now possible to create tiny packages of electronics that contain physical sensors, a wireless router node, and battery. For now, these packages are about the size of a 1-inch cube, but future versions will bring the size down to that of a sugar cube or even smaller. Some forecasters even predict that sensors will become small enough to create something called Smart Paint, which will enable surfaces of rooms, for instance, to sense whether someone is in the room and transmit that information to a nearby computer.

Even at the size of a child's toy block, a self-contained wireless sensor makes a great deal of sense. Dozens or even hundreds of such devices can be situated throughout a refinery or chemical plant, for instance, and collect physical measurements, such as temperature, pressure, or vibration. In the past, each of these sensors would require some kind of wire attached to it to pass information back to a central computer for processing and analysis. But now, with the wireless mesh in place, these nodes can work together to pass information over the airwaves back to the computer. No wires are required except perhaps at the edge of the mesh, where one node could pull in data and pass it into a standard LAN. Every so often, the nodes' batteries might need replacing, but, overall, this approach dramatically lowers the cost of deploying sensors.

Just as on the Internet, this mesh of wireless sensor nodes is self healing: If in the electrically noisy environment of an industrial plant, one node's connection to another gets disrupted by interference, the sending node would simply try to pass data to another node and keep trying until the message gets through. Because only small pieces of data are being sent across the mesh, there's no need for much computing power in each node or much bandwidth in their wireless links. What matters most is reliability, and the redundancy of the multiple links within the mesh offers that in spades. Again, as more nodes are added, the entire network becomes more resilient and useful. There's no central point of control.

Not surprisingly, the ideas behind this technology, sometimes called *ad hoc networking,* have their roots in military research. Armies are as concerned as any corporation with Real World Awareness, especially when the world in question is a battlefield. The military's goal for many years has been to develop tiny wireless sensors that could be sprinkled from an aircraft by the thousands, all across a battlefield. If each one had a magnetometer in it, for instance, capable of detecting the movement of metal objects in its vicinity, this ad hoc mesh of nodes could silently relay the movement of enemy tanks or even individual soldiers. Other sensors might be alert to sounds, or traces of certain chemicals; the possibilities are endless. With enough of these Smart Dust units deployed, of course, the resulting mesh would be hardy enough to withstand the loss of many individual nodes, whether they were run over by a truck or by something else.

The increased power of RFID technology combined with the advances in connectivity are taking the concept of an active tag that can sense and react to its environment to new levels, as the next section demonstrates.

SENSORS AND EMBEDDED SYSTEMS

Although RFID technology can track the precise movement of individual items through even the most involved supply chains, it is often necessary to learn more about an object than simply its physical location at any moment. It may be extremely helpful to find out just how a particular machine is performing right now, or how much fuel it has left in its gas tank, or how many nickels it has left in its change-maker, or how cold its refrigerated compartment is.

This is the technology of remote monitoring and telemetry, or making machines "intelligent" so that they can collect information about their own status and transmit that information to a remote location where someone or some system can interpret the information and take appropriate action. Thanks to the miniaturization of computer circuitry, the falling costs of physical sensors, and the ubiquitous availability of data networks, an era of smart objects and self-analyzing, self-describing, self-reporting machines is close at hand. The Delta Air Lines preventive maintenance scheme for jet aircraft engines, mentioned earlier in this chapter, relies on sensors like the one shown in Figure 3.15.

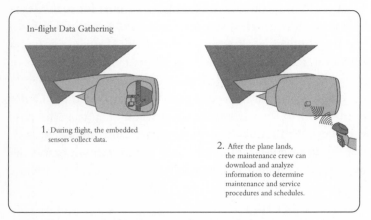

In-flight Data Gathering

1. During flight, the embedded sensors collect data.

2. After the plane lands, the maintenance crew can download and analyze information to determine maintenance and service procedures and schedules.

Figure 3.15. Embedded Systems for Aircraft Maintenance

One simple form this technology will take is an RFID tag that has built into it a sensor of some kind. As production volumes rise and technology advances, the price of tags with built-in sensors is quickly falling, to the point where they are becoming commercially feasible rather than just a laboratory curiosity. These tags have the ability to collect temperature or other physically sensed data and provide it to an RFID reader whenever they are interrogated. Some tags have enough intelligence to compare two readings, separated in time, calculate the difference between them, and send an alert if the difference exceeds some preset limit. The difference in readings, of course, indicates change, and if the change is too much, a dangerous condition may be approaching. Some sensor-equipped RFID tags will be able to actively transmit their findings to a nearby receiver, as soon as they detect a problem situation, and others must wait to be interrogated. Those active tags, naturally, require an on-board battery and will cost substantially more.

One of the most promising applications for sensor-equipped tags is in the area of preventive maintenance. It's often the case that detecting a failing mechanical component early in its decline, well before it has caused real trouble or damage, can save a great deal of money. Noticing that a tire has lost a significant amount of its tread, for instance, can save an automobile owner the enormous trouble of incurring a bad accident on the road. Detecting abnormal vibrations in a pump motor can help the operators of a chemical plant to take action—to fix or replace the motor—before the pump's complete failure causes a daylong shutdown of production.

Increasingly, miniature sensors will be used to monitor machines and their components in ways that were not economically or technically practical before. Tiny sensors that can detect the rotational speed of a truck tire, for instance, can be embedded directly in the rubber of a tire. The sensors can send the data they collect wirelessly to a receiver mounted nearby.

Many automobiles now are peppered with on-board sensors, used to capture all kinds of performance-related information. Some of this data gets used in real-time by the engine itself, and other data gets displayed on the dashboard to keep the driver informed of her vehicle's health. Many cars, meanwhile, can provide still more performance data when they're in the shop for a check-up. Workers simply plug in a specialized computer, and they can view the engine's ignition timing, manifold pressure, and other important parameters. Some new car models capture this kind of information out on the road, analyze it automatically, and, if anything looks wrong, *call the dealer* (via wireless link) to make a service appointment. Finally, many Formula One racing cars now beam floods of real-time performance data to their teams' engineers throughout every twist and turn of a race.

SMART VENDING MACHINES

One of the most telling examples of embedded systems contributing to Real World Awareness is in the new breed of vending machines. In a way, any vending machine is an automated, self-contained retail store, displaying, dispensing, and taking payment for the items it sells. (Figure 3.16 shows how not only vending machines but also shelves and pallets are becoming smart.) It may be useful, therefore, to consider the smart vending machine as conceptually a simple version of the smart shelf that METRO Group and other retailers are developing.

In fact, the vending machine industry is a huge one, and Real World Awareness is becoming a vital competitive weapon there. The purchase of beverages, tobacco products, tickets, and snacks from vending machines is a massive market, particularly in the United States and Japan. Annual revenue from Japan's 5.5 million vending machines is approximately €54 billion. In the United States, annual revenue from 7.4 million vending machines is approximately €32 billion. Vending machine operators are under constant pressure to optimize their processes and integrate them with their enterprise software infrastructure.

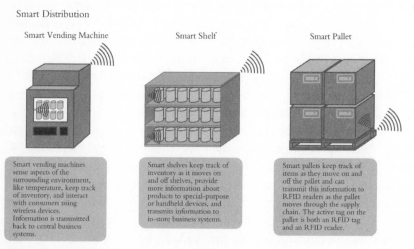

Figure 3.16. Smart Vending, Smart Shelf, Smart Pallet

So, the operators are turning to smart vending machines, which can report their sales activities and other data to a central IT system using a wireless Internet connection. Using information collected this way, operators can better plan the transportation routes for refilling their vending machines and better adapt the range of products to consumer behavior. If a malfunction occurs, the vending machine can send a message directly to a service technician. Furthermore, if a company wants to encourage sales with promotional offers, it may adjust prices on its machines remotely, or update their software and the ads they display.

The messages such machines send can be about almost anything that happens in or to a machine:

"Article taken from compartment 7 on March 12, 2004, at 6.17 p.m. Dollar bill inserted, two quarters given as change. Compartment 5 became empty on March 12, 2004, at 6.17 p.m. Temperature on March 14, 2004, at 2.30 p.m. is 43 degrees Fahrenheit. Attempted break-in on March 15, 2004, at 23.57 p.m."

An *internal,* or *embedded,* computer collects these messages and, depending on the type, urgency, and request involved, forwards them to a central system. The central system evaluates, classifies, and aggregates the messages before forwarding them.

In most cases, the transmission of messages is executed over the cellular phone network. Inside each machine is the circuitry needed to place a cell call to a preset phone number. This method can have a substantial cost, though, as each call gets charged at the prevailing rate. Also, the bandwidth available for moving data on cell nets is fairly low, though this will change as 3G technology comes into play. Eventually, smart vending machines will be able to connect to the Internet via local Wi-Fi networks, wherever they are available, or through longer-range WiMax-based nets. Indeed, the time may come when these machines are essentially on the network all the time and maintaining a pervasive connection day and night.

Such connections will only enhance operators' ability to make more money from the machines they have installed. They can analyze sales data while it is still up-to-date and fine-tune targeted marketing and sales efforts. This data could be information about product sales trends in geographic regions or at different times (such as beer sales are at their highest between 6 p.m. and 11.30 p.m.), or suitable combinations of products (Bounty candy bars do not sell well next to Mars bars, for example). The Real World Awareness connection enables companies to immediately ascertain the effect of advertising campaigns: Do the sales of Coca Cola rise immediately after an ad is broadcast, and how long does this rise last?

That's not all. The analysis of current data from machines can improve the filling process. For example, operators can reduce the quantity of goods they transport because they no longer have to rely on estimates of vending machines' stock levels. Instead, they have fresh and precise information. This ensures that capital is not unnecessarily tied up. Current stock levels can also be used to dynamically plan truck routes. Moreover, this data helps to ensure the availability of products in the vending machines, which is the key to maintaining customer satisfaction and loyalty.

What's more, the ability to send data to vending machines can improve their profitability. Their internal controllers—small computers able to respond to remote commands—enable extensive and precise control. An operator might centrally adjust individual product prices and adjust the temperature depending on the time of day or night. He can block the sale of particular products at certain times. Coffee machine operators can introduce new recipes or alter the blend without an employee having to visit the machine itself.

The benefit of all the sensing we have described is a mountainous flood of information. IT departments will be challenged to manage and assimilate this data into the business systems used to run most companies. Although some of this task will be made easier by the standards mentioned in the following section, the real challenge and the real payoff comes from using the information to respond effectively to business challenges, which is the subject of Chapter 4.

THE ELECTRONIC PRODUCT CODE

In both business and engineering, information technology provides a historically brand-new and extraordinary powerful way of describing the world. Rather than simply use words to describe the world, as writing allowed, or describing events in terms of numbers and formulas, as science and mathematics have made possible, the computer offers a way to simulate the world. Individual objects and their physical and symbolic interactions and relationships can be simulated to any degree of precision. The main limit is time, for the more detailed the simulation, the more time will be required to execute the software that embodies that simulation. As computer hardware becomes faster and new methods are worked out to gang together individual machines, ever more detailed simulations become feasible.

If anything has been learned in the past several decades of using computers to solve business problems, it's the vital importance of standards. When two business partners, or the members of an entire industry, decide to agree on exactly how certain real-world objects or real-world processes ought to be described, they will naturally find it easier to communicate about those items. In fact, what they are doing is making it possible for their respective computer simulations of the real world to overlap and draw information and insight from each other. If Company A's computers identify a certain product as No. 3456 and Company B's computers know it as #AJ-145.R6, there's little chance the two companies can work together with any intimacy. Consider a supply chain with a dozen or more members, and this Tower of Babel problem multiplies geometrically and stifles all chances of collaboration.

As Real World Awareness and RFID emerge, therefore, as powerful new ways of helping companies to coordinate their activities and understand each others' actions in ever more precise detail, the need has arisen for, at the very least, a widely adopted scheme for simply naming all those zillions of physical items of which business computers are rapidly becoming aware. The ability of radio frequency identification to meaningfully identify individual items will be severely limited without a standardized way of sharing RFID tag information (and all other relevant Real World Awareness data) between disparate information systems.

Fortunately for all involved, an open, flexible RFID standard has been proposed and is already in use in many industries, and is under consideration by international standards bodies: the Electronic Product Code (EPC). Building on a decade of research at the M.I.T. Auto-ID center and now managed by the joint venture EPCglobal, the EPC effort attempts to introduce a suite of coherent specifications and protocols governing both RFID hardware and software and the formats of the data they will exchange, store, and process. Of all the standards used in the implementation of Real World Awareness technologies, EPC is perhaps the most crucial. Much as the HTML and HTTP standards enabled the World Wide Web to flourish, EPC is hoped to greatly reduce incompatibilities in the RFID world and help the technology gain widespread acceptance and usage.

The Auto-ID team began with the assumption that RFID would fail to take off commercially until the cost of chips and readers fell dramatically—from 50 cents apiece to less than 5 cents for the former, and from $1,000 to around $100 for the latter. The team also resolved to create a complete standard around which all RFID makers could rally, thus creating economies of scale and driving down costs.

The goal was to create an "Internet of things," in which every object, at the moment of its "birth," or production, could be imprinted with an RFID chip containing an EPC-defined ID number, or code. This EPC code itself is only a string of binary digits—a 64- or 96-bit number (future versions will be 128 bits and 256 bits long). It merely identifies an object uniquely and points to a database server somewhere on the Internet where data about that particular object can be found. Rather than cram data onto the chip itself—which would increase its cost and size—the Auto-ID team's key innovation was to offload that data to networks where storage is cheap, plentiful, and easily accessible. Besides the code itself, the EPC framework includes specifications for tags and readers, as well as for a variety of software needed to manage billions of codes and look up information attached to them, for example. Two key elements are the Electronic Product Code Information Service (EPCIS) and a system named Electronic Product Code Discovery Service (EPCDS), in which enterprise applications look up EPC codes. Figure 3.17 shows you how this lookup service works.

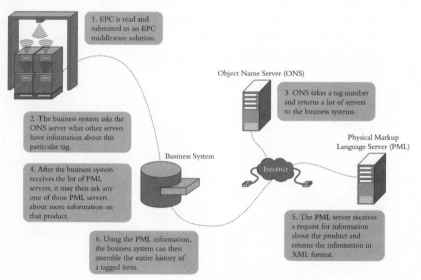

Figure 3.17. EPC Global Standards for Information Retrieval

EPCglobal™ is a joint venture between EAN International and the Uniform Code Council (UCC), founded in autumn 2003. This not-for-profit organization is entrusted by industry to establish and support the Electronic Product Code (EPC) Network as the global standard for immediate, automatic, and accurate identification of any item in the supply chain of any company and in any industry anywhere in the world. Its objective is to drive global adoption and standardization of the EPCglobal Network.

Jonathan Loretto

The EPCglobal Network was developed by the Auto-ID Center, an academic research project at the Massachusetts Institute of Technology (MIT) with labs at five leading research universities around the globe.

Executive Representative

EPCglobal

Jonathan Loretto is the Global Technology Leader for RFID at Capgemini and Capgemini's Executive Representative at EPCglobal. Jonathan, who has a bachelor of science degree in medical sciences, a master of science degree in information systems and management, and a master of business administration degree, is a visiting lecturer at Cambridge University in the United Kingdom.

Q&A with Jonathan Loretto, Capgemini's Executive Representative, EPCglobal

Q: One of the big stumbling blocks to the widespread adoption of RFID is the number of frequencies already in use. Every technology provider seems to prefer a different frequency, and can produce plenty of evidence suggesting that their choice is best. EPCglobal has chosen to rally around the UHF band (902 MHz to 928 MHz), but doesn't UHF have a number of issues?

A: UHF works well in the U.S. because of the freely available space in that range, which allows frequency hopping and phasing to be used, improving performance. But much narrower bands, available in Europe and Asia Pacific, make the use of these techniques virtually impossible.

Q: How so?

A: The UHF band sits in the middle of the GSM band, which happens to be Europe's mobile phone standard. That potentially leads to reader interference if a mobile phone is in use inside an RFID facility. It also means those users are unlikely to obtain a spectrum license in the first place, if that risk exists. They would have to use the frequency on the high edge of the band—928 MHz—which hurts the performance and reliability of the RFID network.

Q: Doesn't UHF have some performance problems already?

A: UHF is a good frequency for dry, medium-to-low-density products, but it isn't good for products containing a lot of metal or liquids. Unfortunately, the laws of physics limit UHF's effectiveness for those things.

Q: Why did EPCglobal choose to go with UHF? Are these drawbacks less than other frequencies? And, is there a way to work around them?

A: The Auto-ID team and EPCglobal chose UHF because of its availability and because the size of its chips can be kept very small—which translates directly to a lower cost for each chip. But it's time to ask whether one size truly fits all. The hardware and software can be tweaked over time to compensate for some issues, but the laws of physics are ultimately intransmutable.

Q: What's a better solution?

A: A multiband RFID system using different frequencies for different products would move the issues surrounding RFID from the realm of physics to the hardware and software arena. A multiband RFID system would consist of multi-frequency, multi-protocol, multi-specification readers and a series of chips designed for different products: A 13.56 kHz tag would handle liquids, a UHF tag for dry goods, a 2.45 GHz one for metallic products, a multi-band tag for complex environments or products, etc.

Q: But wouldn't this increase the costs and the complexity of an RFID solution enormously?

A: Building a multi-band RFID system would be very expensive initially, but the costs would still fall rapidly—there's a need for trillions of units of each kind, and as we move toward actual unit tagging, there will be new, non-silicon based technology —which is necessary if tags are to ever cost less than 7 or 8 cents apiece—able to work better at different frequencies than today's tags. After that, it's just a software issue—a multi-band environment would simply require a software update in order to incorporate new, non-silicon tags.

Another key to this system is that no matter what tag is used, the numbering system must be the same. The essence of any identification scheme like RFID is that everyone can share and understand what a single ID number means.

The Auto-ID team's vision was that every object tagged in this way could be found and have its status tracked remotely as it moved along the supply chain from the moment of its manufacture to the disposal of its packaging by the end user. To that end, the proposed EPC standard also includes specifications for the creation of tiny, low-powered tags capable of containing only EPC codes (thus lowering the cost) and the inexpensive RFID readers needed to read them.

How EPC Works

The EPC itself is a 24-digit *alphanumeric* string (it combines numbers and letters) that can be expressed inside computers as 96 bits of binary data (see Figure 3.18). It alternately functions, depending on your choice of metaphors, as a license plate number, an IP address, or a super bar code designed to do two things: be unique and point to a data file in a computer somewhere on the Internet. This file, sometimes referred to as *metadata,* contains information about the object that the code uniquely identifies. Thus, the code and this server work hand in hand to provide the full capability of the EPCglobal scheme. The EPC code itself contains no explicit information about the object to which it is attached.

Here's how the code works: Its digits are organized in four sections. From left to right, the first section indicates which version of the EPC specification is being used. Next comes a number indicating the corporation that created the tag, which is usually the same organization that produced the item to which the tag is attached. The third section denotes the *object class* of the product: Here, a soda can's tag might carry a number indicating that it's a "Cherry Coke, 12-ounce can, United States version." Finally, a unique nine-digit serial number is assigned to a specific can of Cherry Coke. With nine digits to work with, it's possible to assign unique IDs to no fewer than 999,999,999 different cans.

One of the immediately obvious advantages of the EPC code is its extremely fine level of detail. Although current RFID-based supply chain systems are limited to monitoring the movement of entire containers or pallets and count their contents as a single item, the EPC code allows for every can in every six-pack of Cherry Coke on a pallet to be counted individually. A can going missing can be recognized by the system as a discrete event, which immediately raises red flags. If that level of granular detail proves too demanding, tracking the cans at an aggregated, pallet level can be defined by administrative software. Either way, users enjoy a greater number of options.

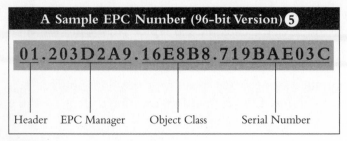

Figure 3.18. An EPC Number

The 96 bits of current EPC codes contain enough unique ID possibilities that, in theory, codes could be attached to every product made for the foreseeable future. There are enough for 268 million companies, 16 million object classes, and 68 billion serial numbers in each class. The future introduction of 128- and 256-bit codes guarantees a surplus of unique codes for years to come. Even if they start running out someday, the code's internal format is such that it can be extended and modified with relative ease, yet without making earlier tags and equipment obsolete. EPC readers would simply be programmed to recognize new codes.

EPC SPECS FOR RFID CHIPS AND READERS

Unlike many other standards, which are narrowly drawn, the EPC standard is unique in that it specifies an entire ecosystem for managing tagged items. The tags and their workings are of course defined, but so are advanced functions of the tags and also software systems to provide ways to find information related to the tags.

The EPC RFID Tag

When the Auto-ID team originally conceptualized what is now the EPC standard, it started from several basic assumptions. RFID tags should be as small and as inexpensive as possible. Mass adoption of RFID, team members reasoned, would occur only when mass production (as measured in the millions and billions of tags) made economic sense for its partners. The team concluded that this would happen when the price of a tag fell to about 5 cents. This initial decision steered the team in the direction of tiny, passive tags that possessed only enough memory to hold the 96-bit EPC code.

The ideal EPC-compliant RFID chip is astoundingly small—only 150 microns across, equivalent to the width of three human hairs. One potential source of such tiny tags: Alien Technology, an RFID tag manufacturer and EPCglobal partner, that has developed a *fluidic self-assembly* process. It uses acid to etch silicon chips at much smaller sizes than is possible with the diamond saws used in traditional semiconductor processes. Alien has shown that its process can produce tens of thousands of tags in a single step. These chips, and others like them, are passive, which means that they are dormant until an RFID reader broadcasts electromagnetic waves at close range (less than ten feet away) and triggers their activation.

Thanks to the reputation of M.I.T. and the Auto-ID team, EPCglobal has collected numerous technology partners working on innovations in chip assembly, antenna design, and materials to create the ideal tags for EPC. But, the pragmatic EPCglobal has kept the standard open enough that any RFID tag operating in the ultrahigh frequency (UHF) spectrum between 902 MHz and 928 MHz and contains a valid EPC is considered standards-compliant.

Different Types of Tags: Class 0, Class 1 Gen 1 and Class 1 Gen 2

The feature sets of the EPC's RFID specs are continually evolving. The two most commonly used now are the original spec, referred to as Class 0, and the more refined spec, called Class 1 Gen 1. The latter, still awaiting approval, adds several new capabilities, including the ability to instruct RFID tags to "kill" themselves.

Class 0 tags are the EPC's vanilla, standard-issue tags. These tags have been given random, non-company-specific EPC codes by their manufacturer and sold for use in a supply chain. It's up to each corporate customer to assign tags to its own physical products and then enter that assignment into a corporate database for tracking. It's useful to think of Class 0 tags as a license plate issued by a manufacturer rather than by the state—the number is unique, but it's up to the driver to associate it with a particular car. Until recently, EPC specs allowed for Class 0 tags to operate at either 13.56 MHz or 915 MHz, although the lower frequency variation has been mothballed for the foreseeable future.

Class 1 Gen 1 tags, also referred to as Class 1.1 tags, can be thought of as license plates issued without numbers, only official markings, thus granting their owners the opportunity to assign their own numbers to the plates. Class 1.1 tags can be written once and read many times via the EEPROM process, which will enable companies to imprint tags with their unique EPC Manager IDs and otherwise map tags to physical products.

Class 1 Gen 2 tags, or *Class 1.2* tags, add the ability to rewrite each tag's contents many times, plus several new protocols that were missing in earlier versions: One is the *kill* feature, in which tags may be ordered to erase their data and burn themselves out in order to prevent future reactivation and reading. Among other uses, this feature may go a long way in reassuring consumers who are wary of RFID tags' potential threat to personal privacy—particularly concerning the idea that government agencies or marketers might surreptitiously scan individuals or their homes to learn about the products people are wearing or have taken home from the store.

The EPC RFID Reader

Although EPC tags have standardized around UHF frequencies, it is now impossible to set a global standard on transmission frequencies for global tags. Because the EPC standard calls for a passive tag, which draws power from radio waves emitted by the reader, the reader itself must conform to international regulations governing electromagnetic spectrum, which are different in the United States, European Union, Asia, and Australia, for example. The solution is *agile readers,* which are capable of scanning tags at different frequencies. These readers now cost upward of $1,000 each, however.

The EPC master plan for an RFID solution assumes a network of agile readers with limited scope, able to scan tags at a distance of about four feet around the 915 MHz and 868 MHz frequencies (the latter is most prevalent in Europe). The Auto-ID team discovered in its field testing that at such a short distance, the weaknesses of UHF—namely, its difficulties with scanning objects through metal or objects possessing a high degree of liquids (Cherry Coke cans would have both)—can be overcome with careful engineering. The Auto-ID team hoped that an agile reader built with its specs might cost only around $100 to manufacture in mass quantities.

OBJECT NAME SERVICE (ONS)

In addition to some specialized middleware that retrieves, reads, and acts on data collected from tags, the EPC effort has defined a specialized database known as the *Object Name Service (ONS)*. The idea behind the ONS is borrowed from the World Wide Web, where URLs typed by users get translated into arcane numerical network addresses before a browser can retrieve a Web page from a remote server. The *Domain Name Service (DNS)* tracks every server connected to the Web and helps browsers find them. The EPC's ONS is quite similar: Whenever an RFID reader scans a tag and passes its EPC up to the RFID middleware, that software asks a local ONS server to point the way to a data file located elsewhere on the network. There, the middleware will find data relating to that EPC code—a detailed description of the item in question, where it's headed, and how much wear and tear it has experienced, for example. At this point, the middleware can grab this data and pass it along to a particular enterprise application that can use it.

To make sure that this data about the object, which may be extremely intricate in structure, is itself in a well-defined format, the EPC group has come up with a markup language for organizing the data. Based on the widely adopted XML standard, this Physical Markup Language *(PML)* ensures that items are defined in terms of well-defined categories. Those cans of Cherry Coke, for example, might be classified as a *carbonated beverage,* which falls under the category *soft drinks,* which is in the broader category of *food.* To ensure that quantities and other characteristics are as uniform as possible, however, PML insists on units of measurement as recognized by the International Bureau of Weights and Measures and the National Institute of Standards and Technology in the United States.

Finally, the EPC framework offers a search facility that applications could use to find all relevant data about a given EPC code. This facility can help determine a code's history as it was passed along the supply chain, for instance.

IMPLICATIONS OF AN EPC-ENABLED WORLD

If EPC succeeds in becoming the industry standard for RFID, it will have not only succeeded the UPC, the standard for bar codes, as the standard for consumer product companies, but it will also have created a much larger sphere of influence and interoperability by bringing additional industries and institutions—including, but not limited to, auto makers, pharmaceutical companies, defense contractors, and even the Department of Homeland Security—into a single, unified system for the first time.

The automotive industry now uses VIN numbers to track parts, and defense manufacturers use the NATO standard STANNAG for their components. Pharmaceutical corporations have their own systems for the development and manufacture of drugs. The EPC framework now exists as a nexus point for these disparate fields, in which institutions and governing bodies, such as the U.S. Food and Drug Administration and Department of Defense, have issued mandates calling for a move to RFID. The EPC contains the promise of not only succeeding the UPC but also becoming the unifying standard for formerly industry-specific systems.

From a technology perspective, EPC is perhaps the most ambitious standard ever attempted, with hardware, software, and metadata standards already in place. If EPCglobal succeeds in its evangelist efforts and wins the full embrace of (at least) consumer product companies, it will have prevented potentially years of battling over proprietary standards, billions of dollars in dead-end development efforts, and billions more in savings for adopters by creating economies of scale.

This chapter has covered lots of ground. The technologies involved in sensing the real world have been explained along with the connectivity and standards that greatly ease design and deployment. Information for its own sake is not enough, however. To derive value from Real World Awareness, a useful response must be crafted that takes advantage of the information that is gathered. The response is also assisted by technology, as it must be to manage the huge volumes of data and the automated relationships that are used to create value in Real World Awareness deployments. In Chapter 4, we look at how businesses will use the information that is collected and delivered by the technologies described in this chapter to change they way they operate and compete.

4

Business Process Design and Optimization

It was a momentous day in the history of business innovation when in early 2003, Wal-Mart, the U.S. retailing giant, formally asked its top 100 suppliers to start adding Radio-Frequency IDentification (RFID) tags to all the goods they ship to its distribution centers and stores. Although Wal-Mart was not requiring the use of RFID tags until early 2005, its decree immediately increased the importance of RFID for every industry. Suddenly, this technology that had been invented decades ago and had been moving ever so slowly into the marketplace was given a big push. And, ever since, it has been gaining momentum. The retail sector was the most obvious and immediate beneficiary, but every other sector is now on notice that RFID will soon be playing an important role there, too.

But what role, exactly? Although it was clear how RFID tagging would help Wal-Mart and other organizations that were mandating its use, including the Department of Defense (DoD) and Boeing Aircraft, executives at those top 100 retailers and at hundreds of other companies might be forgiven for not immediately grasping how RFID might benefit their operations, too. To comply with the Wal-Mart mandate, for instance, a supplier would likely invest millions of dollars in purchasing RFID tags, setting up new IT infrastructure to oversee the data within those tags, and institute new manufacturing processes for attaching the tags to individual packages and products. Then what?

This chapter attempts to answer that question as fully as possible, by using real examples of RFID adding value and enabling the rethinking and revamping of important business processes in a number of different industries. This chapter demonstrates that RFID and Real World Awareness can be used to make incremental improvements and to foster deep, long-lasting business innovations.

In a way, history was repeating itself when Wal-Mart insisted on its suppliers using RFID. Certainly, this was not the first time that a Wal-Mart technology decree had created such a sense of urgency. Years before, Wal-Mart had called on major suppliers to begin using bar codes in new ways, and their positive response had quickly made intensive bar code usage standard operating procedure in industry after industry. Now, the Wal-Mart call for the use of RFID is having a similar effect. And, important questions are starting to be asked across corporate America and within overseas companies: "What will RFID mean to us? How can *we* benefit from this technology, too? How can RFID help us innovate in our own business? Which of our business processes can gain the most from this technology? What new business models do Real World Awareness make possible?"

The obvious benefit of RFID for any company directly affected by the Wal-Mart mandate is that compliance will help ensure their survival as a supplier to the world's largest retailer. Beyond the obvious sort of "slap and ship" scenario—slap on a tag and ship the product—the technology has the potential to provide tremendous value to corporations: It will help companies squeeze inefficiencies from their logistics operations and give them better visibility into their supply chains. RFID will also help companies leverage real-time information about operations, aided by a menu of analytic tools, and help them improve their replenishment processes. In brief, RFID technology will help companies create and profit from the so-called consumer-driven supply network, with which they can be more responsive to market conditions and changing demand.

Consider this statement: Until fairly recently, enterprise computing was concerned mainly with recording and executing business transactions—orders, shipments, and payments, for example. More recently, methods were developed for delving into masses of transaction records—usually well after the fact—to identify patterns and correlations and thereby gain new insight into customers and their behavior.

Companies, then, could use these insights to create more accurate fore-casts, develop improved products, and fine-tune their marketing and promotional activities. Naturally, as computers gained speed and storage capacity, this processing and analyzing of transactions was greatly improved, which led to new business efficiency and effectiveness. Add low-cost, ubiquitous data networking, and entire business processes could be reworked, automated more thoroughly than ever, and in some cases speeded up by orders of magnitude.

For example, although it might have taken days, in the 1970s, for a catalog merchant like Sears Roebuck to process an order arriving by mail, Web-based retailers now process and confirm customer's orders within seconds. And, whereas Sears published a new catalog every few months, at best, a company like Amazon.com can adjust the products, prices, and marketing messages it presents to individual customers in milliseconds, as quickly as any visitor to its Web site clicks from one Web page to another.

Real World Awareness is poised to shorten and accelerate even further the vital feedback loops that connect the world out there and the world—or, more accurately, the symbolic model of the world—inside the computer. Now, rather than merely enable customers to serve as data-entry clerks—which has been, in fact, the Web's primary contribution to retailing—Real World Awareness turns potentially every machine and object in the world into a data-entry clerk. *Every thing,* in other words, becomes able to sense some aspect of itself or its surroundings and, more or less immediately, communicate those findings to the appropriate business systems. Rapidly disappearing is the need for workers on load-ing docks or in factories, for instance, to manually enter data through keyboards or to scan bar codes to make note of the tasks they have just accomplished or record observations they have made.

Thus, with no intervention from any human, a warehouse manage-ment system can be kept aware of exactly where inside a facility each pallet of goods under its control is located at any moment. A refriger-ated delivery truck can report not only how many cases of ice cream are on board but also the temperature of its cold compartment—and per-haps its location, highway speed, and even the amount of wear on each of its tires. A jet engine may collect data about its turbine's in-flight per-formance, by describing rotational speed, temperature, and vibrations. A vending machine at a remote highway stop can communicate to its owner, far away (perhaps by wireless network) that its inventory of Mars bars is getting low. A "smart shelf" in a retail store can report each time a customer removes a product—or puts it back. And so forth.

The wiring of the world with so many sensors, of so many types, all collecting and reporting data about themselves and about the goings-on in their vicinities, has profound implications for businesses and for the managers who run them. If nothing else, this wiring of the world demands new investment by companies in IT infrastructure, ranging from truckloads of RFID sensors to networks of RFID readers to wireless data networks and more. A whole new tier of software is required, too, to collect, filter, and distribute the massive volumes of sensor data that will start flowing into corporate data centers. New techniques and tools will be needed to analyze and make sense of these floods, too.

OPPORTUNITIES ABOUND

Real World Awareness is not solely a matter of shouldering new burdens, however. It opens the door to myriad new opportunities for innovation, both in how businesses operate and how they are managed. Many opportunities will arise to incrementally improve well-honed operations and also to rethink, from end to end, certain key business processes: Wherever there is a physical movement of objects—merchandise in a distribution center, products moving through an assembly line, delivery trucks on the streets and highways, or high-value diagnostic equipment in a hospital, for example—there's a good chance that Real World Awareness can help improve operations.

And, for those with a particularly creative bent (and no small amount of good luck), Real World Awareness may serve as the basis for lucrative new lines of business, new business models, and entirely new companies. This set of technologies may not prove as fertile as those that ushered in the World Wide Web and attracted a torrent of investment by venture capitalists, but Real World Awareness is sure to stimulate a great deal of new thinking, innovation, and hard-dollar payback. Consider just this simple, albeit unproven, example:

While browsing in a large retail store, a shopper finds a television set that interests her. To make sure that she's getting a good price, she pulls out her cell phone, which is equipped to read the RFID tags attached to individual products. While the shopper is receiving data by a price-comparison service to which she has subscribed, she thumbs through a listing of the prices that other merchants—located online as well as down the street—are charging for the model of TV she wants. Satisfied that the store she's visiting is asking a good price, she decides to buy the TV there.

How much would such a service appeal to consumers? How many would sign up? At what monthly charge? Might retailers feel threatened and try to block this kind of service by insulating their stores against cell phone signals? Who knows? Just as it was impossible in, say, 1996 to foresee the vast range of new dot-com businesses that were about to spring to life on the World Wide Web, so is it virtually impossible right now, this early in the Real World Awareness game, to predict the variety of new ideas and business models that may be spawned. Many of them will be trivial and short-lived, but some, it's a good bet, will survive long enough to establish themselves as thriving enterprises. And a few, you can be sure, will be hands-down winners—the next Amazon.com or eBay, perhaps.

Preparing for the Challenge

Indeed, by this point, the incorporation of Real World Awareness into corporate computing looks inevitable. It's unstoppable, for various reasons, and that means that the challenge for every company right now is to figure out how to think about and apply these technologies in ways that deliver maximum return on investment. Like the Internet, which burst onto the business scene in the mid-1990s and sent corporations around the world scrambling, this new wave of sensor, communications, and analytic technologies cannot be ignored except at the risk of failure. As Charles Darwin might say, were he alive to ponder the relentless and accelerating evolution of business technology, corporations have just one simple choice to make: Adapt or die.

For many companies, the pressure to adapt is quite explicit. In several sectors, certain organizations whose size and buying power make them impossible to ignore or resist are mandating the use of radio frequency identification technology by all their suppliers. Mass retailers such as Albertson's, Target, and Wal-Mart in the United States and Tesco and METRO Group in Europe are starting to insist on the use of RFID by thousands of consumer goods manufacturers. As of late fall 2004, the Wal-Mart plan was to have its top 100 suppliers tag their shipments with RFID by early 2005. Next, the company would require that suppliers begin using those tags to improve the accuracy of their advanced shipment notifications (ASNs), which inform Wal-Mart about how many goods have been loaded onto each truck that's headed its way.

That's not all. Boeing and Airbus are also insisting that their suppliers use RFID, as a way to better manage the flow of raw materials and preassembled parts arriving at massive aircraft assembly plants. Finally, the biggest mandate of all is the one issued by the U.S. Department of Defense, which buys many billions of dollars worth of goods each year. With RFID tags on everything from trucks to blankets and cases of ammunition, DoD logistics managers will be better able to track the minute-by-minute movement of material and more quickly reroute supplies to any place on the globe where they're needed.

Depending on their size, suppliers affected by these mandates will opt for different means to achieve compliance, much as occurred years ago when major manufacturers such as General Motors and Boeing promoted a computer-to-computer business-communications scheme named electronic data interchange (EDI). Smaller suppliers stand a good chance of being able to "make do" with what are essentially prepackaged setups. These suppliers, working with outside consultants, IT vendors, and specialized systems integrators, should be able to pull together the necessary hardware and software to add RFID tags to all the goods they ship to Wal-Mart or METRO Group. Ideally, though, these firms would also extend their own IT systems and modify their own business processes in ways that could enable them to directly benefit from RFID use, too—perhaps as a method for tracking finished-goods inventory and recording each shipment as it leaves for a particular customer.

Larger companies, whether responding to another firm's mandate or acting on their own, will likely opt to assemble their own Real World Awareness setups from a combination of off-the-shelf components and custom-designed software. This way, they will gain the advantage of systems that precisely meet their needs and goals and fully support their most critical business processes. For most large companies, Real World Awareness can be added to existing IT systems—as a replacement for bar code scanning, for instance—without a great deal of effort in terms of reprogramming and thereby quickly improve their operations. Only a small amount of modification to their IT systems will be required. Later, to achieve the full benefit, these companies may see the need to undertake a more thorough modification and rethinking of business processes and underlying IT systems, and that will require substantial planning and preparation—as is the case with adapting to any important new technology.

The METRO Group of companies comprises the world's third largest retail concern. Its autonomous divisions—including METRO AG, Metro Cash & Carry, Real, and Extra, among others—reported combined sales of EUR 53 billion in 2003. The METRO Group now employs more than 250,000 people at 2,406 locations in 28 countries.

Zygmunt Mierdorf

Chief Information Officer

METRO Group

As the Chief Information Officer of the METRO Group, Zygmunt Mierdorf is in charge of the company's human resources and social affairs, IT, logistics, real estate, and e-business initatives, including the groundbreaking Future Store. This ongoing experiment in next-generation retailing technologies is a collaboration with SAP, Intel, and IBM, in addition to other technology companies. Before becoming chief information officer in 1999, Mierdorf was managing director and spokesman of the Management Board of different companies in the METRO Group from 1991 until 1998. Born in Frankfurt in 1952, Mierdorf studied economics at Wiesbaden Poly-technic and graduated in business administration.

Q&A with Zygmunt Mierdorf, Chief Information Officer, METRO Group

Q: What's the true potential of RFID for a major retailer? Cost cutting? Enhancing the customer experience? Competitive advantage? What's METRO's real motivation?

A: All three of those issues. RFID will bring significant cost savings or efficiency due to process streamlining. The customer can also expect great benefits because the systems will take care of replenishment automatically. If we really realize our vision of paying without the corre-sponding payment operation (that is, just pushing the shopping cart with the products through a reader gate and then that amount is recorded and debited), it's a significant time-saver and simplifier for the customer. And, of course, we can use RFID for providing additional informa-tion for the customer, not just at the terminals, but also via screens. Whenever the customer moves around within the store, wherever we have installed this technology, we are able to pro-vide information and specifically target him.

Q: What about competitive advantage?

A: The competitive advantage is relatively clear. If the shopping event becomes more adven-turous and more attractive, we assume that one benefit is customer retention. It definitely offers a competitive advantage if one views it with regard to cost; and if one has implemented it in a meaningful way, one can utilize the savings achieved in marketing or wherever one wants to utilize it to filter through to the bottom line.

Q: What's the most interesting avenue of RFID development right now? What are you excited about?

A: The most promising issues at the moment are logistics, due to the reason that everything else is possible only if one has 100 percent coverage with RFID chips. We are starting with the simple things. We will then broaden this to product tagging and then apply theft protection on the tag. That will be removed at checkout and is rewritable, so it can be reused.

The issue of automatic replenishment is next, and the in-store systems depend on the speed of RFID spreading in the consumer goods industry, because these systems work in a meaningful way only if they are area-wide and 100 percent available. But we should try to get the out-of-stock problem under control through RFID—that is a significant problem, for the retail side as well as for the supplier side, because that means lost revenues.

These will take place successively—logistics is first, and then security, and then we will combine them in the tag. And then in a third phase, the customer will see benefits in product availability and checkout. But those are relatively long-term processes; there, we are talking about a period of 10 to 12 to 15 years.

Q: I think the brand managers of these products are more worried about out-of-stock than you. If Procter & Gamble has no toothpaste on the shelf, but there is one from Colgate next to it, who has the bigger problem now, P&G or you?

A: We both have a problem. One can always argue that the customer will grab an alternate product when he has gotten used to certain products and does not find them, and that happens repeatedly, then we are creating customer disappointment. The direct impact is surely at the manufacturer, but if that happens often, it may lead to the customer changing his point of sale, and then we have a problem as well.

Q: What will be the biggest mistake made by retailers with RFID technology?

A: The biggest mistake—and we must all be aware of it—is over-promising. That is, don't set expectations too high and don't be lured by the press into defining dream scenarios here because they will backfire.

The second big mistake would be to implement it shrouded in secrecy. That means this technology must be increasingly accompanied by a professional marketing and communication policy. That is the most important thing, to eliminate the basic fears of the consumers of a violation of their privacy. One must make everything one is doing there extraordinarily transparent—inform the customer about what we do, why we do it, and where we have these tags and then give him the opportunity to deactivate them, to take them down, or to do whatever else.

THINK PROCESS, NOT TECHNOLOGY

As always, the very real danger is that managers themselves will become seduced by the new technologies and then appreciate them only superficially and misunderstand their true purpose and potential. Indeed, a full, in-depth evaluation and understanding of the business processes that make a company as efficient and competitive as it is will be mandatory for any except the most superficial implementation of Real World Awareness.

This seemingly obvious advice is borne out by research conducted by PRTM and SAP, based on a survey of some 60 companies and 75 different supply chains. The conclusions: Companies with mature business processes have significantly lower inventory levels—28 percent lower, in fact—compared to immature companies. Companies that invest solely in improving business processes leave money on the table, no doubt, but those that invest only in IT infrastructure end up being far more inefficient than they should. Indeed, without mature business processes, even companies with the top 20 percent, best-in-class IT setups were found to have 26 percent higher days of supply, 28 percent higher inventory-carrying costs, and 7 percent lower profit than those that were process immature and did not invest in IT. Evidently, implementing just IT systems without their supporting business processes is a waste of money.

THE APPROACHING FLOOD

Now, of course, bar-coded pallets and packages are tracked by the millions as they make their long journeys from factories to retail shelves and, finally, through the laser beams of checkout scanners. So, what is it that RFID might add to this well-documented success story? A great deal, in fact, including much improved visibility into inventory, quicker detection and prevention of theft, accelerated supply chain speed, and more meaningful exchanges of information between business partners.

RFID data will be better data than what has been available from bar codes and other sources. It will be *more accurate* because of its being created automatically, with no manual keyboarding, and it will be *collected more frequently* than what is now collected from bar codes. As a result, this data will be richer in informational content and will therefore enable IT systems to maintain a *more detailed representation, or model, of the world.*

Real World Awareness technologies in general and RFID in particular will cause a major leap in the sheer volume of digital data with which enterprises will have to cope. Because RFID tags can be scanned automatically, with no human intervention, each of them will likely get read many more times during its active life than any bar code attached to the same product or other object. RFID readers are getting to be so inexpensive that enterprises can install them in just about any point in a supply chain where they want to observe the movement of items—at any door or loading dock, in any truck or store room, or in the hands of any worker with a wirelessly connected personal digital assistant (PDA), for example. RFID readers will show up in many more places than has been the case with bar code scanners. Eventually, individual shelves in a supermarket—*smart shelves*—may have their own readers, too.

Along with greater frequency, there will be an increase in the complexity of data—an increase in the categories of data that will be collected and available for transactional and business intelligence systems to work with. One way to think of this concept is that RFID data will arrive from the field in a *more granular form* than what is now commonly obtained. Traditionally, consumer goods have been tracked by the pallet, case, and individual item, but those objects have generally been classified quite crudely—by brand, type of good, and quantity. A bar code may indicate, for instance, that its underlying object is a case of 12-ounce boxes of Kellogg's corn flakes. With RFID, however, each and every case, and every box of corn flakes, may be given its own, unique ID number, thus enabling each one to be individually tracked from factory floor to checkout counter. When a pallet of cereal boxes is loaded onto a truck, not only may the pallet's tag be read but also the tags attached to each case of boxes and all those attached to cereal boxes. Thanks to the EPCglobal numbering scheme, as described in Chapter 3, each of these packages may be assigned a unique serial number that distinguishes it from every other item in the universe.

The downside of all this? More data-collection points, more frequent collection, and finer-grained data will result in a veritable flood of new information that, left unchecked, will overwhelm IT systems as they now exist. Rewriting those systems to handle all this new information is out of the question, so what's needed is a new layer of software that can shield enterprise resource planning (ERP) and other key systems from the coming flood. As supplied by SAP to its ERP customers, this vital new software layer goes by the name of Auto-ID.

The Auto-ID infrastructure performs several key functions, as shown in Figure 4.1. At its core, it maintains a master record of all the RFID tag numbers—or, more precisely, the EPC numbers recorded in specific tags—that an enterprise is keeping track of. And, this software maintains a record of the relationships that exist between all the objects those tags identify. For example, Boxes 176 through 226, which contain apple pies, are stored on pallet No. 23-4B, which was last scanned as it entered Section A-92 of Warehouse 314159. To keep up with the constant movement of the many items under its purview, this software must operate at high speed, to constantly update its record of items and their locations and also provide selected snapshots of this data to any business systems that request them. As designed by SAP, a copy of the Auto-ID software will run in each store, warehouse, and other facility in which a customer has installed RFID scanners. These multiple instances of the software are linked, however, in such a way that they act as a single repository of information.

One important task for this new programming will be to ensure the integrity of the data it is capturing. That means cleaning up and normalizing data as it flows in from the field. One frequent problem that early RFID setups have revealed is that of duplicated messages. When a pallet, for instance, passes within range of an RFID reader, any information systems working with data from that reader should receive only a single reading of the pallet's tag. As it turns out, RFID readers are designed to interrogate all nearby tags once every second or so, which means that if the pallet gets stalled in front of the reader for some reason—too many forklifts trying to get through an exit door, for instance—it might generate hundreds of identical, and therefore confusing, signals before it moves on.

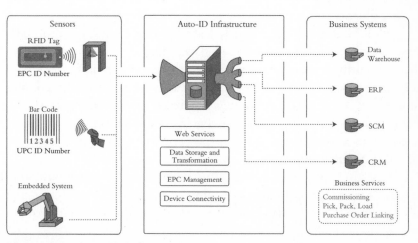

Figure 4.1. Auto ID Infrastructure

Another important task for the Auto-ID software is to facilitate access to its data at the precise level of detail, or *granularity,* that any particular enterprise application may require. Some systems will work from long lists of highly detailed scanning records, and others will need high-level summaries. In either case, the Auto-ID infrastructure's job is to provide the freshest possible data through well-defined interfaces and service calls. Only in this way can users be sure that any new applications they develop will be able to partake of RFID's bounty.

ANALYTICS

With this new software in place and devoted to storing and serving up masses of sensor-based data, an important question arises: What next? What kinds of tools can be applied to the data to distill new information from it and to unearth the most valuable nuggets of insight hidden in its depths? The answer is a variety of analytical tools, some of which already exist and some of which must be developed specifically for use with this new kind of data.

In some cases, flesh-and-blood managers sitting at keyboards will do the analysis, as they do now with point-of-sale (POS) data and other transactional information. In other cases, IT systems will get programmed to do at least some analysis and interpretation autonomously, to identify patterns and relationships that are ripe for exploitation. Predefined business rules may get triggered for execution whenever certain thresholds are crossed—an impending out-of-stock condition in a heavily promoted soft drink, for example—and business managers alerted by pager or cell phone when a situation requires their immediate attention.

Let's look at the kinds of analytical tools that can be used to make sense, so to speak, of sensor-based data. Brand managers and supply chain executives cannot be expected to be geniuses in mathematics; they need tools that can help them explore and make sense of the coming volumes of Real World Awareness information in a straightforward, down-to-earth manner. Luckily, two genres of such software are available to corporations for just this purpose: online analytic processing, or OLAP, and data mining.

These tools are by no means entirely new; they have been available in various forms for more than a decade. But they can and will be applied in some powerful new ways to help deal with the real-time nature of the data that Real World Awareness technologies are starting to produce. What these tools share in common is the ability to help analysts deal with rich, multidimensional data. This is data whose individual records (each one describing the details of a particular transaction, for instance, such as the purchase of an item in a store or, in the RFID realm, the passage of a pallet or other item from one location to another) contain many fields of data. In brief, the more fields of data in a record, the richer is that record and the richer and more potent is any collection of such records as a whole.

Consider a POS record as a simple example—the kind of record that a retailer's checkout terminal might generate. This record might capture only the item's price, the time of its purchase, and the store where the purchase took place. Clearly, by sifting through a large collection of such records, it would be possible to determine what kinds of items a particular store sold the most of during each hour of the day. This information might prove useful in planning the movement of inventory onto particular shelves, just to keep them from running out of stock.

Now, imagine if each POS record also captured data about all the other items that each customer bought during a visit to the store. A pattern might be detected: Customers tend to buy beer and corn chips at the same time, for instance. This insight could be used to place those items closer together in the store—or perhaps further apart, as a way to encourage customers to make longer journeys across the retail floor and increase the chances that their eye would be caught by additional items. Each retailer might interpret and act on these results in its own manner, based on its unique merchandising strategies.

Expand the POS record even further and things get quite interesting: Each record might also include data about a customer's gender, demographics (income bracket or education, for example), and previous purchases at the store. And, with RFID-based smart shelves involved, data could be obtained about products that each specific customer removed and later returned to a shelf—an event that might be interpreted as indicating strong interest in the product, which, with some properly formulated promotional efforts, might well be pushed along to become an actual purchase.

The more data that is available about individual events and different aspects of each customer's behavior during a store visit, the more useful the information that can be inferred about that customer's interests, desires, motivations, and habits. And, the more of that kind of insight that's derived from the mass of raw data records, the more effective the store can be with its advertising, promotional, and merchandising efforts. This concept is simply common sense.

What may be less obvious is that as these data records become individually richer, and comprise more independent fields of data, considerably more effort is required in order to identify all the most useful and meaningful patterns and correlations that may exist across a collection of such records. Indeed, the required effort increases geometrically, if not exponentially with each increase in data complexity. When dozens of variables, or fields of data, are involved, simple human intuition no longer suffices, and neither does merely plotting a couple of variables for visual analysis. Instead, some powerful, computation-based techniques must be employed, techniques that can explore what are, as mathematicians view them, multidimensional spaces that the human mind can grasp or navigate only with extreme difficulty, if at all.

Let's look at each of these analytic techniques and see how they will help cope with the fruits of Real World Awareness.

ONLINE ANALYTIC PROCESSING

Increasingly, as computers have been able to capture more data about more business transactions of various kinds and recorded more pieces of data about each individual transaction, tremendous volumes of data have become available for exploration. Specialized repositories called data warehouses excel at collecting, normalizing, organizing, and keeping at close hand these volumes of records and making them available for rapid analysis.

Imagine a marketing specialist, for instance, who may want to explore a mass of historical sales transactions in hopes of determining the three or four factors that seem to most strongly indicate who is likely to buy a certain product—a certain kind of shampoo. Once these factors are identified, of course, they can be used to fine-tune a marketing campaign. This analyst may have a hunch that it's suburban mothers, age 28 through 45, with three children or more, and household incomes of $100,000 or more. The question is, how does the analyst go about proving her conjecture, based on all this hard data that's available?

The answer is a technique called online analytic processing, or OLAP. It is a set of software tools that work on masses of multidimensional data—data whose individual records may include perhaps 100 or more independent variables. To make sure that a search through 500,000 complex records takes only a few seconds, the records and their many data fields are organized in a special way, with rich cross-indexing paths linking their fields in mathematically precise and efficient ways. This organization of the records is quite different from what's found in the more common relational database, whose job is mainly to locate one or a small set of records in a flash and perhaps update them with 100 percent reliability. Think of OLAP processing as involving tasks such as looking for the intersection of two sets of records, each numbering 20,000 and with 50 subfields of data per record.

Using OLAP, our marketing specialist can quickly see how important each of the variables she has selected may be, and she can see what weight any chosen combination of variables may have. If her hunch is wrong, she may choose some other variables to explore. Each time, the OLAP system can zip through the necessary searches and produce answers within seconds. Doing a single such search on a standard relational database, in contrast, might take hours. OLAP software is getting significantly faster as computer hardware advances and makes it possible for more information to be kept in the fastest random-access memory (RAM). What's more, as seen in an OLAP product such as SAP's Euclid, some preprocessing and aggregating of data can greatly speed the search process, too.

DATA MINING

In some cases, a mountain of rich, complex data may exist but the people who would like to explore it have only the vaguest ideas about where to look for patterns and relationships. There may be too many fields or variables to contemplate and intuition is of no help. This is a job for data mining technology, which is able to find correlations and patterns in such data more or less on its own.

That may sound like magic, but it's not. The trick is to apply sophisticated statistical and machine-learning methods, which together can sniff out "facts" that are hardly obvious even to those quite familiar with the data and the "world" it supposedly describes. With little or no pretraining or guidance, data mining techniques can zero in on clusters of records that share some set of characteristics. Perfected over the past decade or so, data mining has been a fairly esoteric branch of analytics, requiring a certain amount of training on the part of users. But increasingly, the technology is being commercialized in forms that are easier to use and more productive in a shorter amount of time.

And with RFID just around the corner, the need to further refine and apply data mining techniques is gaining a palpable urgency. These tools look like the best means available to fully exploit the value locked away in the masses of data that soon will be pouring into data centers from the sensors of all kinds. Analysts will have to learn to trust their data mining tools, even if they don't fully understand how those tools find the answers they do. Once they are in use, data mining tools will help retailers to explore now-unseen aspects of their supply chains and of customer behavior. For example, with RFID scans available in large volumes, it will be possible to evaluate the performance of logistics providers as never before. It may turn out that one trucking company, though charging more money than others, actually excels in making rush deliveries but not standard ones. Another firm, meanwhile, might show by far the best value—measured in terms of on-time performance versus price—for non-rush deliveries. Such comparisons would not be possible without evaluating a raft of empirical data collected from many actual deliveries over a long period of time.

Likewise, within a retail chain, data mining techniques might be used to determine why a certain product sells better in some stores than others. After evaluating a range of possible variables, the data mining algorithm may determine that the most significant influence is where on the retail shelf a certain competing product is placed. Or, it may find that a particular combination of promotion and customer demographic is clicking in one store location but not another. So many possible variables and combinations of influences are at work in these situations that it's virtually impossible for any human analyst to filter out the important patterns and effects on his or her own.

RESPONSIVE REPLENISHMENT

Even as improved general-purpose analytics become available, specialized algorithms are also being refined to help make the most of RFID and other Real World Awareness data. Perhaps the biggest payoffs, so far, can be envisioned in the area of inventory management, especially in mass retailing. By providing vastly improved information about where goods are, precisely, at any moment—in this distribution center, in the back of that truck, on shelf No. 17 in store X—and by doing away with the errors that occur when data is entered by keyboards, RFID will greatly improve retailers' ability to make the right goods available exactly where and when customers want to buy them. They will be able to quickly react and adapt to fast-changing business conditions and make their supply chains more consumer- and demand-driven than ever before.

Indeed, better information will help foster significant improvements in existing processes and operations as well as provide the foundation for true business innovation. Scanning RFID tags on objects throughout a supply chain will squeeze inefficiencies from such processes as warehouse management and logistics. Freight shipments will be traceable with much better accuracy and speed. Because RFID data will feed into ERP systems, just about every process those systems help automate will benefit.

As for true innovation, there has been tremendous technologic effort focused on the inventory-management problem for many years, particularly in the consumer packaged goods arena. There, improving the processes of forecasting, planning, and replenishment is critical to continued competitiveness and success. There is perhaps nothing worse in the consumer products business than succeeding with a marketing campaign and attracting many customers to a store, only to disappoint them by failing to provide the very items that they have come to purchase. Not only are they more likely to switch to a competitor's product for that purchase, their loyalty to the original brand and to the store they are visiting may be significantly diminished. Out of stocks, as these situations are called in the trade, are to be avoided at almost any cost.

Easily one of the biggest innovations in replenishment has been the notion of vendor-managed inventory (VMI). Conceived some two decades ago, VMI calls for a rich exchange of information between retailers and their suppliers—a constantly operating feedback loop that works like this: The retailer provides fresh POS data to the supplier, describing how many units of each item have been purchased over, say, the previous week or couple of days. The supplier now uses this information to plan deliveries of additional items to the retailer's warehouse or individual stores, striving to maintain the inventory there within a predetermined range of minimum and maximum levels. By reducing the amount of extra inventory that would otherwise be kept on hand to prevent a shortfall—so-called safety stock—this scheme can save the retailer significant money. And, it gives the vendor more accurate insight into how its goods are selling, perhaps in response to a new marketing campaign or in the face of a competitor's recent advertising activity.

In short, there is a move from "push" to "pull," from manufacturers shipping products pretty much as it suits them to retailers working together with manufacturers to closely monitor consumer demand and ship only as much as is actually needed. This is a fundamental shift in supply chain philosophy and one that is affecting both "mass value" supply chains and those centered on "new luxury" goods. The former are managed with an aim of maintaining high production volumes and keeping costs down. The latter, operating on the principle of "sense and respond," focus on adapting to fleeting consumer fads and fashion trends as quickly as they appear and disappear.

One of the weakest aspects of current VMI setups is that the sales activity data retailers send to vendors arrives in batch form after some non-trivial delay in time. Although sales take place minute by minute during the business day, sales data gets aggregated over a full day, or a week, and then gets delivered in one fell swoop. The result is a lag between the moment when purchases get made and the moment when a vendor can decide how best to react, whether that's by shipping more goods or perhaps even boosting production of certain items. To compensate, the vendor—because he is operating under the threat of penalties should the inventory he controls fall below a certain threshold—may ship more goods than are actually required, just to be on the safe side. But because of how the VMI relationship is typically structured, he will be the one who pays for those extra goods—indirectly, through the expense of financing them while they wait to be sold. The vendor may also misjudge a spike in sales and start producing more goods than will actually be consumed.

This tendency of suppliers to over- and underreact while trying to understand fleeting shifts in demand is referred to in supply chain circles as the bullwhip effect: A small flick of a bullwhip's handle can cause the end of the whip to move very far and very fast—faster, even, than the speed of sound, thus creating a small sonic boom, which is what produces the whip's distinctive "crack." And coming up with ways to dampen this over-amplification of signals and the accompanying over-compensation has been a major endeavor ever since the CPG and auto industries, for instance, began adopting VMI. Ideally, each sale of an item would trigger a corresponding replenishment of the shelf from which it was taken—though for obvious reasons, such a fine-grained, 1-for-1 replacement is quite impractical. Still, much improvement has been attained, with many leading CPG suppliers and retailers now successfully working together on daily replenishment cycles.

Daily replenishment, executed well, might seem quite sufficient and perhaps even verging on overkill. But manufacturers such as Procter & Gamble and Gillette and retailers such as Wal-Mart, METRO Group, and Tesco are actively searching for ways of managing retail inventory on a sub-daily basis, using information technology to tighten the feed-back loop even more. And here is where RFID will truly shine.

What's driving the quest for sub-daily replenishment is the fact that the CPG marketplace is increasingly driven by frequent and aggressive promotional activities initiated both by suppliers and retailers. Indeed, more than 50 percent of the business transacted is now being driven by storewide sales, holiday sales, advertising campaigns in newspapers and on television, special offers on selected items, cents-off coupons, and similar promotions, all aimed either at getting consumers to try a par-ticular brand or product for the first time or at simply attracting shop-pers to stores. It's virtually impossible these days to launch a new cosmetic or food product, for instance, without the help of heavy pro-motions. Consumers need to be prodded to try new things, which is why CPG makers spend as much as 18 percent of their sales on pro-motions. And with more retail outlets selling a broad range of goods—and in many cases essentially the same goods at more or less the same prices—there is fierce competition over getting consumers to choose one store over another. And just to make the situation more complex, suppliers and retailers often run promotions independently of each other, for their own purposes, without even forewarning the other of their plans.

What this increase in the frequency and range of promotional activ-ity means for supply chains is much greater volatility and much less pre-dictability concerning demand—and therefore the need for greater agility in trying to match supplies to that extra-volatile demand. Even the most successful promotion will be defeated by a lack of goods in the right store and in the right quantities, ready to be purchased by newly motivated consumers. But with various promotions, including those initiated by

competing suppliers and retailers, affecting demand in different ways practically every day, managing inventory becomes a mounting challenge.
Should a particular item start, one day, to move off the shelf more quickly
than expected, both retailer and supplier have an interest in keeping that
shelf replenished as effectively as possible. If the store's back room starts to
run out of the item, the vendor would benefit a great deal from being
notified of that situation as quickly as possible. Likewise, when a promotion is observed to be falling short of its intended effect, it's helpful to be
able to act quickly and attenuate shipments of a product.

RFID promises to help out by providing more precise information
about where additional supplies of goods are in a supply chain so that
they can be moved quickly to where they are needed most. This, in a
nutshell, is the demand-driven supply network, offering a fluidity and
agility not possible before. With data from RFID tags showing exactly
where goods are, it becomes easier to improve the performance of the
entire supply chain.

As P&G has discovered, making responsive replenishment work
requires some new thinking in the area of forecasting and planning
algorithms. The software used now has been designed to handle only
daily updates about sales activity, at best, but with more data flooding in
from RFID tags on a sub-daily basis, it becomes necessary to modify the
software. In the past, forecasts of how many items will get purchased
have been made on a weekly basis, with the total amount divided
equally across each day of the week. With sub-daily information available, it becomes possible to profile and forecast sales activities separately
for each day of the week, taking into account such constraints as the
store's opening and closing time and the delivery schedules of different
vendors' trucks. All this means more detail and more complexity, and
planning algorithms will need to be modified to keep up.

RFID will have additional benefits for the consumer products business. In general, the superior data provided by RFID will tighten up
supply chains, eliminating slack and "wiggle room." For instance, it's
common now for retailers to deduct 5 percent to 10 percent from each
invoice they receive for goods that a supplier has shipped to them. The
retailers have learned that they can get away with this deduction by
claiming that the amount of goods received has fallen short of what they
had ordered. The supplier is forced to investigate and come up with
proof of some kind that the right number of items was shipped and
delivered. In fact, it is quite possible that the number of pallets a supplier

ships does not match the number that finally arrive at the retailer's warehouse: Pallets may be miscounted as they enter a truck, data entered into a keyboard may be inaccurate, and pallets occasionally get stolen from trucks en route to their destinations. But with RFID readers automatically recording the movement of pallets and doing so at many more locations along the way from factory to store shelf, there is much better data available about exactly what gets put on each truck and what leaves it. RFID readers can even be installed inside trucks, to periodically monitor their contents and beam the information wirelessly to a central location.

As new planning algorithms come on stream at companies like P&G, successfully improving their ability to fine-tune replenishment, the effects will be felt far up the supply chain. There will be opportunities to shorten production runs, for instance. Today, these runs tend to be relatively long, designed to minimize product costs and the costs of changing over production machinery from making one kind of product to making another. But with finer-grained demand data available, it may be possible to shorten those runs in a way that is optimized to demand. That, of course, will require the development of new metrics, or criteria for measuring manufacturing's performance. New metrics may give more weight to a production facility's flexibility than in the past, when sheer throughput, or production volume, was valued most.

ADAPTIVE MANUFACTURING

The term used to describe this kind of highly flexible production is adaptive manufacturing. And in pursuit of that flexibility, many corporations are striving to organize and participate in so-called adaptive business networks. In a nutshell, this idea calls for manufacturers and their suppliers to stay in touch with each other through enhanced information sharing and be able, therefore, to quickly react to demand signals received from the marketplace—directly from consumers, as might happen with web or phone-based orders, or from retailers selling their products. The poster boy for this kind of setup is Dell Computer, which doesn't start building any computer before actually receiving an order for it. Indeed, by maintaining intimate links with its suppliers, Dell is able to avoid ordering certain components from them before receiving an order for the PC that will use those components. This last-minute ordering further enables Dell to pay for those components later than is usual in most industries. With lower inventories of components and raw materials on hand and the ability to pay for components as late as it does, Dell has been able to achieve levels of profitability that are the envy of its rivals.

In general, the trend in supply chain management has been to move from a "push" model to a "pull" model. Manufacturers seek to build or produce goods only as their customers ask for or buy them. In the past, manufacturers tended to push products down the supply chain with inventories in warehouses and stores helping to buffer supply against changing demand. Now, with the pull model, it's necessary to make manufacturing much more adaptive, or agile, so that goods can be built more or less as needed. This way, the factory acts as the supply chain's main buffer and inventories, which cost money to finance, are kept to a minimum.

The consumer packaged goods industry has been especially keen to adopt the pull model and it has spent a great deal of money on developing the necessary technologies. This has been a challenge, of course, for products such as cosmetics and foodstuffs because they aren't produced quite the same way as a Dell PC. Indeed, these kinds of products are usually produced in batches—so many thousands of gallons of a certain type of shampoo, say—which are then packaged and shipped down the supply chain. But even here, unforeseen demand for that particular shampoo product, perhaps stimulated by a retailer's promotion, may catch the manufacturer off guard. He may not have enough labels, for instance, to produce a sufficient number of bottles for the shampoo. With the right information links in place, however, he would be able to quickly work with his packaging supplier to get a rush order of the right bottles. So, even in what have been largely batch-production industries, agility and adaptability are increasingly important.

And that is where Real World Awareness is proving to be a big help. By RFID-tagging raw materials, components, sub-assemblies, and work-in-progress (WIP) as they enter a production line, for instance, it becomes possible to maintain a real-time model of exactly where in the production process each important element of work is at any moment. Then, if a particular tool breaks down, managers can quickly determine the state of the production process and make informed decisions about how to work around the breakdown and keep things moving. Without the help of RFID, they would have to rely on intuition, lots of telephone calls, and perhaps even manual methods such as blackboards and hand signals across the factory floor.

Likewise, most machine tools already produce a good deal of potentially useful information which, if collected and interpreted properly, could provide another powerful way to monitor production operations in real-time. Sensors are in place to measure the mechanical performance of the tools themselves, and other sensors are capturing the movement of items on assembly lines and so forth. One obstacle to making good use of this data has been the lack of a standard format. Each tool maker has done things its own way, more or less. But lately, great strides have been made in working out a standard for shop-floor integration, as it's called, and in rallying support for this standard from leading machine-tool makers.

And now, perhaps for the first time, manufacturers can achieve what SAP calls top-floor-to-shop-floor-integration: Upon collecting near-real-time demand signals from the marketplace, production can be rapidly tailored to meet that demand. And, if there's a problem in one production facility, work-arounds can be quickly devised, perhaps calling other facilities into action or lining up new suppliers of a suddenly scarce component. In any case, the ultimate goal is agility, with Real World Awareness helping to manage things at the level of physical activity. Figure 4.2 shows the general structure that enables this responsiveness.

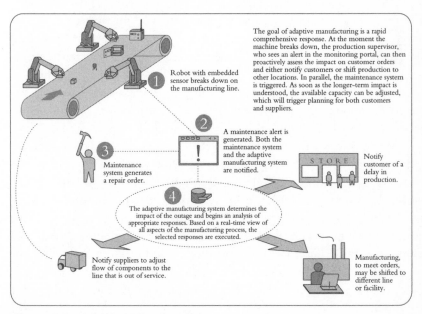

Figure 4.2. Adaptive Manufacturing

TRACKING PRODUCTION AT INTERNATIONAL PAPER

A good, early example of how RFID can help the manufacturing process can be seen at International Paper (IP). It's using RFID technology to achieve real-time visibility of assets and WIP material. The company's RFID setup has been running for more than a year at a Texarkana, Texas, bleach-board mill, tracking the stocking and storage of thousands of rolls of paper each day. Since implementing the system, the Texarkana mill has increased inventory turns by 5 percent, improved on-time delivery performance, and leveraged the system's preplanning and tracking capabilities to ship 70,000 tons of paper in March 2004, a new record.

The new technique of RFID has helped to turn warehousing into a function that actually helps improve the production flow. To understand what's produced and stored in the Texarkana mill is to understand why IP picked that location for the implementation of its first RFID solution. The mill first converts raw timber into massive rolls of so-called bleach board, which gets used to make milk and juice boxes. The products in the Texarkana mill look like huge spools of thread stacked on top of each other, in endless rows. These rolls, weighing between two and seven tons each, will build up a large sea of white paper.

Forklift trucks outfitted with padded hands grip and lift the rolls and stack them 3 to 8 feet high, depending on their weights. In this environment, IP is facing a unique storage and tracking situation. As the warehouse must move more than 5,000 rolls a day, product tracking and placement has always been a challenge. Attempts to reorganize the warehouse using painted lines and three generations of bar coding lead to unsatisfying results in terms of inventory accuracy, so they were abandoned. Searching for a solution, IP's Smart Packaging business unit decided to test RFID applications newly being developed in-house.

The director of the Smart Packaging business unit, Steve Van Fleet, had been involved in early Electronic Product Code work at the Massachusetts Institute of Technology Auto-ID center. When he saw Motorola using electrically conductive inks to print computer-readable information on its products, the idea came up that IP might use the same technology to run its packaging. The vision: A tracking system that would use RFID tags incorporating Electronic Product Code standards to give each roll of bleach board a unique identifier while providing visibility into how it is processed, where and how it is stored, and how it is shipped.

Challenges and Solutions

The project's performance specifications were set up by IP, but the technology required partnering agreements with hardware, software, and systems-integration vendors already active in RFID system design. One of them was MATRICS, of Rockville, Maryland, a maker of tags and readers. The latter have been mounted on the padded forklift hands that pick up and move the rolls. APRISO was chosen to provide software that maps out locations and directs forklift drivers via truck-mounted computer screens, and finally it was decided to work with ESYNC for the systems integration and consulting.

The project was not without its challenges. First, the development team struggled to find the proper radio frequencies for transmitting information inside the mill. The second big challenge was designing brand-new technology that could withstand the harsh surroundings. The requirement was to ensure that the setup is able to read through a 75-inch diameter roll of paper and still survive in the mill environment. To make that possible, IP and its vendors developed rugged, gripper-mounted readers and placed the tags inside the cores of the paper rolls, a move that significantly reduced the chance of their getting damaged.

With the new RFID scheme up and running, it's possible for the mill's information system to hold all the specifications for each roll, including the customer's name, date needed, grade, width, and mode of transportation entered at order entry. That data is linked to a tag, which receives a unique identifying number and is applied to the roll's core. After production, the roll is picked up by a forklift using grippers equipped with tag readers that give the driver detailed information about that roll, including the next set of instructions concerning where to move the roll. If a roll is to be shipped out by truck that day, the screen displays which conveyor system it needs to be on and whether it's going by truck or rail—and the product has to be handled only once. If a roll needs to be stored, drivers are directed to preplanned storage locations organized according to product range and diameter.

Talking about Revolution

The Texarkana team claims that there isn't a process or a procedure that has not been revolutionized by the implementation of IP's RFID system. One of the biggest changes is that the mill is now capable of tracking a constantly shifting population of rolls inside the warehouse, which means as many as 5,400 individual moves each day for product already in storage. Because each tag includes very detailed information, the system also supports the mill in planning operations and managing transportation costs.

To maintain efficiency and engage employees in the project, the Texarkana team developed a program loosely based on the Quarterback Rating (QBR) in the National Football League. The QBR rates passers against a fixed performance standard, based on the percentage of completions, average yards gained per attempt, percentage of successful touchdown passes, and percentage of interceptions. Texarkana's version of QBR rates forklift operators on the number of valid, invalid, and efficient moves, as well as average moves per day. An invalid move into the wrong conveyance, for example, would be equivalent to an interception; the successful loading of 30 or 40 rolls into the correct conveyance, leading to a complete and on-time shipment, would be a touchdown.

If a driver has taken a roll to the wrong conveyance or truck door, the system immediately issues an alert and the move is noted on the driver's score. A correct move improves the driver's rating. But the program does more than simply motivate employees: It is a real-time, high-level application that gives a record of everything a driver has done for that particular day.

By every measure, the Texarkana RFID project has been a notable success. But Texarkana won't be alone in that success for long: IP's Smart Packaging business unit is now rolling out its warehouse tracking system at several other warehouses and mills. What's more, IP is now looking outside the paper industry for potential customers for its unique system.

PREVENTIVE MAINTENANCE

Potentially one of the most powerful classes of algorithm that can help make sense of and respond to sensor-generated data is something called *predictive analytics*. Techniques have been developed for periodically collecting data from various kinds of machinery while it is operating and analyzing the data for minute patterns that point to malfunctions. By identifying these problems well before a machine breaks down, operators have the luxury of arranging for necessary repairs on a schedule that suits them, at much less cost and with much less disruption than if they waited for a failure to occur.

Deutsche Post World Net (DPWN) is one of the world's leading logistics companies. Comprised of its Deutsche Post, DHL, and Postbank brands, DPWN serves more than five million business customers in half of the world's Fortune Global 500 companies. The Deutsche Post division delivers mail to 39 million German households. DPWN has more than 380,000 employees, and in 2003 reported revenues of EUR 40 billion.

Klaus Zumwinkel

CEO and Chairman of the Board of Management

Deutsche Post World Net

Dr. Klaus Zumwinkel is CEO and Chairman of the Board of Management for DPWN. He rose through the ranks of McKinsey from 1974 until 1984, acheiving the rank of senior partner before leaving to become chairman of the board at, first, the Quelle group in 1985 and, later, at Deutsche Bundespost in 1990. Born in 1943, Dr. Zumwinkel has degrees from Muenster University and the Wharton Business School.

Q&A with Klaus Zumwinkel, CEO and Chairman of the Board, Deutsche Post World Net

Q: What areas of research are DPWN focusing on in its RFID trials? What are the advantages in using RFID in supply chains, transportation, and warehouse management? What are the advantages in using RFID to deliver letters and parcels?

A: The goals of our current trials are twofold. On the one hand, we have to check the maturity of the technology for use in challenging logistics environments. Physical problems have to be overcome, for example, the interference of RFID systems caused by metal or fluids in their surroundings. The read ranges and read rates of transponders need to be improved. The bulk reading of up to about 50 items per second is a very promising feature, but it needs further development, too.

On the other hand, we're exploiting how it might influence our business. Currently we use barcodes to identify shipments. What do we need to do to prepare for the new amount and quality of data that RFID can generate? And RFID applications might also affect our own processes—we can now think about some kind of "self check-in" of shipments.

The overall benefit we expect from RFID is better visibility of the transport chain and using this information to optimize our network and processes. RFID offers a better data quality due to the reduction of manual errors. Bulk reading might allow us to check if parcels on a pallet are complete. With more expensive transponders we can offer functionalities like temperature monitoring.

In the mail business, however, due to the high degree of automation and cost-efficient barcode systems, the use of RFID on every letter is not viable. Nevertheless, even in the mail sector for some closed loop applications RFID offers commercial value.

Q: Assuming the trials go well, how will RFID change DPWN in the long run? What is the best case scenario?

A: RFID will not fundamentally change the logistics industry. But RFID is a promising technology for the enhanced identification of things. Examples of this might be inbound control, improved inventory management, picking control, increased track and trace for perishables, advanced security solutions for high value goods, or anything else.

Assuming our trials go well, in the next phase we can embed RFID in our Express business. We're thinking about using RFID to improve the internal processes of a global network infrastructure and support near real-time track and trace. Today's supply chains consist of many partners like shippers, airports, airfreight carriers, etc. RFID will allow easier integration of all partners into an extensive logistics system.

Q: How does the cost of RFID tags affect DPWN's plans for using them? And what other challenges must be addressed before widespread adoption occurs?

A: For a global, cross-company solution that comes close to the idea of a universal identifier—usable not only by us internally, but also by our customers, shippers and receivers—the prices for transponders and hardware must go down. Only then does a full rollout make sense. But there is no fixed cost barrier.

In open networks, international standards are indispensable for all involved and to see the real-time status of the shipped goods. As a leading global logistics service provider, DPWN is all the more interested in global standards as we serve companies from very different industries in countries all over the globe.

Q: How will you protect customer privacy in such a network?

A: We have to deal actively with these concerns by showing the benefits of RFID and also its technical limitations. Many of our employees—and the public at large—are already users of RFID, and view it as an essential component of daily life.

Building and vehicle access control, car security systems, libraries, etc. are just a small number of examples where RFID technology is widely used by the general public. But RFID is not omnipotent. It has physical limitations like limited read ranges of a few meters, so "Big Brother" visions are not based in facts. Often the privacy discussion is not RFID-related but mixed with the problem of preventing the misuse of personal data stored in databases. We strongly hope that the public discussion about chances and risks of RFID will be realistic.

One of the most illuminating examples of predictive analytics at work is at Delta Air Lines, which applies it to the maintenance of jet engines. Figure 4.3 shows the general structure of preventive maintenance of jet engines. Getting an early heads-up about problems developing within any particular engine enables Delta to schedule and make repairs in a way that entails the least cost and disruption for itself and its customers. In the past, the airline scheduled maintenance work and replacement of engines based solely on the number of miles each one had flown. That turned out to be a crude and ultimately inefficient measure of engine wear. Now, each engine's performance is continually monitored for early signs of trouble, and repairs are scheduled to preempt serious trouble.

The worst case for Delta, or any airline, is that an engine suddenly develops a problem in mid-flight and its pilot is forced to shut it down and perhaps even land the plane at an unscheduled airport. This situation can cost Delta as much as $1.5 million. Only slightly less troublesome is when an engine proves unusable just before a flight is to begin, thus grounding the plane and possibly leaving customers stranded. If the airline is lucky, a spare engine or the required parts will be available at that airport, but too often a replacement engine must be shipped by truck from a distant location. Between that engine shipment, one plane left unusable for a day or two, the need to rush a replacement plane into service, and several hundred disgruntled passengers, the costs to Delta are not insignificant. Any forewarning about engine problems could yield big savings and also help Delta keep its assets as productive as possible and its customers happy.

Having wrestled with this problem for years, as has every airline, Delta turned to the start-up company SmartSignal for help. SmartSignal has commercialized an algorithm that Argonne National Laboratory originally developed for monitoring the health of components within nuclear power plants. The algorithm can detect telltale patterns in the kind of "noisy" performance data that all machines and mechanical systems generate. Because no machine is perfect, temperatures, the rotating speeds of axles, voltage levels, the frequencies of vibrations, and even ambient noise all tend to vary slightly moment by moment, and hour by hour, even when all is okay. The challenge that the SmartSignal software addresses, therefore, is how to filter out those natural fluctuations and "zero in" on any movements in that data that indicate nascent problems.

The software does this by using some advanced statistical correlation techniques that constantly compare the latest batch of sensor data to a "model" that's based on past sets of data that are known to be normal. About 50 "snapshots" of data from each engine are needed to determine this model.

A typical Delta jet engine has 12 to 15 sensors that measure temperatures here and there within the turbine, the turbine's speed, the flows of air and fuel, and more. The sensors are there as part of the engine's design, intended to alert pilots to immediate problems that may arise during flight. But SmartSignal captures the sensors' data for its own purposes, by collecting "snapshots" of all the sensors' readings at selected moments during each flight. Not surprisingly, a jet's engines are under the most stress during take-off, so the first data snapshot gets captured at the moment the airliner reaches an altitude of 300 feet. Additional snapshots are taken each hour, after the plane settles into its cruising altitude and speed. Each time, the data is passed by radio link up to a satellite and then beamed back down to earth and through the Internet to the SmartSignal computer center. There, computers immediately analyze the new data in search of anomalies.

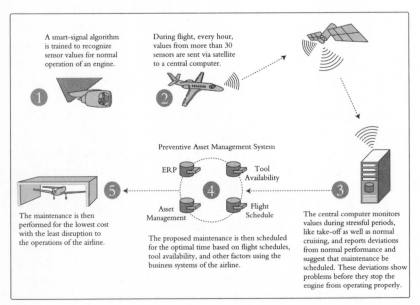

Figure 4.3. Preventive Maintenance

Delta Air Lines, with revenues in 2003 of $13 billion, is one of the world's largest air transportation companies for passengers and freight. It serves more than 200 U.S. cities and 50 international cities in 32 countries. Delta and its partners now operate 7,800 flights each day to nearly 500 cities in 87 countries. The company, based in Atlanta, employs more than 60,000 people.

Ray Valeika

Senior Vice President for Technical Operations

Delta Air Lines

Ray Valeika was Senior Vice President for Technical Operations at Delta before his retirement in October 2004. Before joining Delta in 1994, he was Senior Vice President of Technical Operations for Continental Airlines and, before that, Vice President for Maintenance and Engineering at Pan American Airlines. He served as chairman of the Air Transport Association (ATA), the Aging Aircraft Task Force, and chaired the revision that introduced damage tolerance criteria into the development of maintenance programs. He is a 35-year airline maintenance veteran. Valeika graduated from St. Louis University, Parks College, with a bachelor's degree in aeronautical engineering.

Q&A with Ray Valeika, Senior Vice President for Technical Operations, Delta Air Lines

Q: How can RFID and Real World Awareness technologies affect engineering, systems integration, and other areas in order to ultimately increase safety?

A: Over time, there has been great progress in the cockpit, but really no progress of the same level in how maintenance is performed. In the 40 years that I've been in this business, we still pretty much do the same things, and that is beginning to change because of technology and integrated RFID.

Maintenance is about information. And most maintenance issues are unplanned, they tend to be random. With more information, you can eliminate random events. Lastly, it starts moving maintenance away from the craftsman mentality toward a more information-driven mentality. And I think that assures a lot more safety because you're not depending on one person's judgment. You're moving judgment toward a broader base of people, and the opportunity to find knowledge in that base is much greater than just an individual.

The first big breakthrough for RFID will be real-time knowledge about the equipment on the airplane. The second, and this is particularly important with RFID, is "What is the configuration of that part relative to its maintenance program?" And the third is "What is the configuration of that airplane relative to the reliability of that part?" So, now you know the parts that are there, now you know each part relative to its requirements, and you also know the part's medical file. Through RFID, you can connect not only the part number and where it stands in the maintenance program, but also the whole history of that part and where it stands relatively to reliability. Is it a part that fails frequently? Infrequently? Et cetera.

Q: Will the application of RFID lead to evolutionary changes or revolutionary ones? What will it make possible, rather than refine?

A: The next big change in maintenance will be to make the airplane talk to you, much more so than it does today, and so I see a variety of possible sensors we haven't even thought about. With an engine today, we monitor some very basic things—fuel flow, speeds, rotation, etc. The next phase is the application of sensor technology. I think we should have optical sensors; we should have sensors that can smell. With an engine, for instance, if it were burning fuel the wrong way—maybe there is more carbon than usual or maybe more sulfur—it could smell that.

Or, take the boroscope. Right now, I use my eyeballs. Why couldn't I digitally X-ray an airplane the day it is built and then, each time it comes in, take another set—just as doctor would of your body—and see if there's a change?

Today, we're primitive. We depend on eyeballs. Tomorrow, I think you will have airplanes that will have tremendous amount of sensors. There is sensor technology on a new fighter airplane that would absolutely astound you, and it has not been applied mainly because we don't know what to do with the information and how to get the information.

Over time, it would start to remove human error on the ground as well. Today, we have a $150 million machine that sits on the ground, waiting for somebody with an eyeball to look under a galley to see if there is corrosion. That's insane. Tomorrow, I will take an engine and change only one blade instead of removing all the parts in the engine. We spend time on airplanes taking parts off to see if it really is the part that's supposed to be there, those kind of things. And so a lot of the unscheduled maintenance will be prevented by just understanding those three configurations I talked about: the parts, the reliability of a part, and where a part is in the maintenance program.

The SmartSignal algorithm is sensitive enough to detect incredibly minute problems in these engines. In one case, the software detected a suspicious vibration, and when members of the Delta repair crew tore the engine apart, they found that the culprit was a quarter-inch crack in one of the turbine's fan blades. Over time, that crack would likely grow longer until the blade disintegrated and perhaps wrecked a good part of the entire turbine. In another case, a slightly high rotational speed was detected in an engine. The engine's manufacturer said that this speed-up was well within the acceptable range. But, later, while doing routine maintenance on the engine, the Delta repair crew found that the engine had sucked in a bird, whose carcass had been blocking a temperature probe. The engine, as it was supposed to, had increased its speed slightly to compensate, but the effect was to burn excessive fuel.

When a red flag goes up on a certain engine, the Delta repair people can start to make plans to fix it, which is where predictive analytics truly pays off. Working with a maintenance and spare-parts scheduling system, managers can determine when the relevant plane will next be on the ground and free to be worked on at a repair facility that has the right people, parts, and available capacity. In less urgent cases, the plane may already be scheduled for maintenance work while, in others, a special appointment may be required. Either way, Delta saves money by getting the work done at the optimal moment.

The SmartSignal software has found success in a variety of industries. Entergy, a utility company, is using it to monitor machines—pumps, boilers, and motors, for instance—in 33 power plants. Every 10 minutes, the SmartSignal data center receives readings from 17,000 sensors installed there. Eventually, predictive analytics may find use on static structures like bridges and in trucks, cars, air-conditioning systems, and even home refrigerators. And, in some industries, the SmartSignal software may serve as the basis of entirely new lines of business: A company monitoring its own assets might do the same, for a fee, for other players in its industry.

SERVICE MANAGEMENT

One of the most critical—and profitable—tasks for many manufacturers these days is providing post-sales services. These can range from consulting services aimed at helping customers figure out how best to use a complex product all the way to the business of maintaining and repairing products out in the field. Until recently, many manufacturers considered maintenance and repair as a sort of necessary evil, an expense that they would just as soon have customers shoulder themselves. But now, service has become a major profit center, and companies are eager to find ways of maximizing their income there.

Not surprisingly, RFID can play a key role in this regard, for the service-delivery supply chain suffers from many of the same information-related problems and challenges that show up in traditional supply chains moving goods from factories to warehouses to retail stores. For instance, service-delivery supply chains need to track service assets as they are moved from location to location. They must keep supply matched to unanticipated variations in demand. And such supply chains need to maintain the highest possible utilization of service assets—the trucks, the tools, and the people that make up the service-delivery supply chain.

In this section, we look at how RFID is helping in a pair of service-related scenarios. The first example is Airbus, the highly successful European airplane manufacturer, and the second is Fraport, the company that operates Frankfurt Airport.

AIRBUS

Airbus doesn't just build many of the world's best-selling and most widely flown aircraft—it helps its customers to keep those airplanes in tip-top shape, to fly as often and as safely as possible. Helping out with selected portions of the aircraft service task means that customers (airlines such as United Airlines and Lufthansa) can make more money with their Airbus aircraft, and it means that Airbus can add to its bottom line.

One of the most important support services that Airbus provides is the rental of a range of highly specialized tools that aircraft repair crews need when working on certain components and parts of their aircraft. Airbus has chosen RFID as a way to help automate and improve its processes for leasing and maintaining these tools and tracking them throughout their life cycles, as they get selected, used, and returned by customers and then, with great precision, recalibrated by Airbus personnel. Airbus has begun attaching RFID tags to the tools as a way to record data about each one's technical state as well as certain administrative information. Airbus aims, by using this data, to make its supply chain more transparent and increase each tool's availability and utilization. These tools are quite expensive, and each minute that they're not in use is a minute of lost revenue.

Deutsche Lufthansa AG is one of the world's leading airlines. Alongside its core business of passenger and freight transport, Lufthansa offers its customers a number of specialized services related to air travel. More than 90,000 employees from 150 countries have made Lufthansa not only one of the world's leading aviation companies, but also an employer with a particularly international focus. Lufthansa is among Germany's 20 largest and most popular employers.

Stefan Lauer

Executive Board Member
Deutsche Lufthansa AG

Stefan Lauer has been a member of the Executive Board of Deutsche Lufthansa AG since August 1, 2000. He now heads the Group division Aviation Services and Human Resources and is simultaneously Labour Director. He is responsible for the strategic management of the Logistics, Maintenance, Repair and Overhaul (MRO), Catering and IT Services business segments, as well as for the traffic regions China and India. Stefan Lauer joined the Lufthansa Executive Board on May 1, 2000 as a deputy member. Previously, from January 1, 1997, he had been a member of the Executive Board of Lufthansa Cargo with responsibility for Marketing and Sales, and latterly Executive Board Chairman.

Q&A with Stefan Lauer, Executive Board Member,
Deutsche Lufthansa AG

Q: How do you think RFID technology will affect Lufthansa?

A: As an aviation company, we see ourselves as the motor of a mobile society. We will use RFID sensor technology to make our service safer, more reliable, and more efficient. In our networked society, it is essential to network business processes intelligently. Technology such as RFID helps us optimize processes and offer high-quality services.

Lufthansa Group's business segments are in discussions with logistics partners and leading aircraft manufacturers who are promoting RFID technology. Innovation is one of Lufthansa's trademarks, alongside quality, safety, and reliability. That's why we are open to technological advancement and carefully examine the potential benefit to our customers. With FlyNet, for example, we became the first airline to offer Internet access onboard.

Q: What potential does Lufthansa see in RFID, and what are the requirements for harnessing this potential?

A: RFID is now enabling a transition from the "virtual world," which is based on assumptions, into a "real world" in which objects, such as an air-freight container become "visible," thanks to their ability to be "seen" by using radio signals.

This technology, which allows automatic and contact-free communication, has broken through barriers that previous technological optimization could not overcome. We see RFID's primary uses in building security, asset management, maintenance, supply change management, and logistics—areas in which its application will give customers additional value—for example, in tracking baggage, recording mandatory maintenance work, and locating cargo pallets.

Our management philosophy—as decentralized as possible, but as centralized as necessary—also applies to our approach to RFID. Each individual Lufthansa business segment judges RFID's value in its field of business and launches its own, independent project, which accelerates the overall process chain while saving costs.

Q: In which fields of business do you see concrete uses for RFID?

A: Lufthansa is currently examining RFID in several different projects. Our logistics partners and aircraft manufacturers, in particular, are happy to collaborate with us in order to accelerate the implementation of new processes.

Conventional technologies, such as the bar code, always need line-of-sight vision to the device reading them. RFID systems do not require this and put us in a position to realize new savings potential. By using this technology, we can automate our processes and make them more intelligent and more transparent for our customers.

From my bird's-eye view of the company, I see great potential in networking all types of maintenance and logistics processes.

An example of this is Lufthansa Cargo's work on the IATA's (International Air Transport Association) Cargo 2000 initiative. As part of this initiative, Lufthansa Cargo and LICON (Logistic Ident Consortium) launched Project Laurel to construct an intercontinental RFID supply chain. Alongside the goal of testing RFID technology on the Munich/New York route, Lufthansa Cargo is also focusing on establishing a global RFID supply chain standard. Only by having established standards can airlines and logistics partners use innovative logistics solutions to bring additional value to the customer.

Lufthansa Technik Logistik (LTL) is investigating using RFID to track replacement parts in its warehouse, its goods issue department, and inbound delivery. There is a much larger potential for savings in using RFID technology throughout the material supply chain, however. LTL is in discussions with aircraft manufacturers and Lufthansa's maintenance, repair, and overhaul department.

In addition, LTL has assigned Lufthansa's IT services department with the task of constructing a GPS-based tracking and tracing system for the stands used to prop up aircraft engines during maintenance. Better traceability and secure dispatching will ensure that equipment is used more effectively by shortening the amount of time it spends in transit, which allows the company to operate with smaller numbers of special engine stands and saves both LTL and its customers money.

The general goal here is to make internal logistics processes more transparent, accelerate processes, and minimize errors across all Lufthansa business segments.

Q: Are industrial standards necessary for coordinated implementation of RFID in the aviation industry?

A: Industrial standards and standardized processes are most certainly necessary. Industry solutions should be given preference over customized solutions wherever possible. The technical requirements in terms of the frequencies used across the globe need to be established at government level and the issue of data protection resolved explicitly.

RFID is now in a phase of hype and inflated vision—for example, the vision that everyday objects will be made "visible" by the radio communication ability that RFID will give them. Despite this phase, we expect RFID to have a significant impact on our business processes in the long-term.

For this reason, we support all activities that would bring us closer to an industrial standard for international airlines. A standardized view of process chains could be advantageous to all companies involved, which is why we are also working closely with our logistics partners, aircraft manufacturers, and airports.

If we manage to get everybody together at one table, we will be able to accelerate implementation, particularly because networking processes across companies will lead to higher return on investment than if each company looks only at its own processes.

The Airbus customer service unit, known as the Spares Support and Services division, is based in Hamburg, Germany, and operates service centers in Washington, D.C., Singapore, and Beijing. It operates 365 days a year and 24 hours a day and employs 490 people, of which 400 work in Hamburg. Revenue in 2003 was €111 million. The division has 11,500 tools in stock, of which more than 4,000 have been equipped with RFID tags. RFID is nothing new for the Hamburg facility, for it has been using the technology for the past three years. However, the tags used there, only eight millimeters in diameter, have contained only limited technical information about each tool: part and serial number, weight, volume, and date of manufacture.

At the start of 2004, Airbus began to make the RFID tag information available in the supply chain process, by setting up an information channel between its SAP system and the RFID tag readers.

FRAPORT

Fire safety is a top priority in all facilities that serve the public, but perhaps nowhere more so than at an operation as large and as busy as Germany's Frankfurt Airport. It serves millions of passengers every year. Fraport AG, the company that operates this airport, has embarked on thoroughly revamping its maintenance of fire-safety equipment: ventilating-unit fire shutters, security doors, and smoke alarms, for instance. Until recently, the regular, scheduled maintenance of these many devices was handled with a highly paper-intensive process. RFID tags are now helping to automate much of the work and thereby eliminate paper, cut costs, improve accuracy, and help maintenance crews operate more efficiently, as shown in Figure 4.4.

At the end of 2002, Fraport began planning a mobile system for maintaining the more than 22,000 fire shutters that exist in the airport's 420 buildings. Until then, workers were assigned maintenance work via paper printouts, and when they finished their work, they were required to fill in more paper forms describing the work they had done, any parts they may have replaced, and so forth. The result was not only a sea of paper forms, many of them filled in manually with inadvertent errors, but also the temptation for workers to simply avoid doing the work they were supposed to. Without a follow-up inspection of each set of shutters, airport management had no way to positively know that any maintenance work had been done.

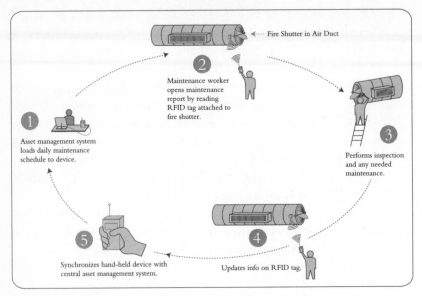

Figure 4.4. Mobile Asset Maintenance

The Fraport response was to develop a computer- and RFID-based application that eliminates reams of paperwork *and* ensures that maintenance work gets done as it is supposed to. Using SAP software, the new system puts a handheld computer in each work team's hands. At the start of each workday, this computer is briefly connected to the Fraport main maintenance system, which downloads to the "handheld" a list of work orders: the specific repairs to be done on specific shutter installations.

These orders are not fully accessible to the workers yet, however. All they're able to see is the location of the first shutter they should work on. When a worker arrives at that shutter, an RFID reader connected to a handheld collects information from a tag previously attached to the shutter. If this tag's ID number matches what the handheld expects to find, the full work order is unlocked and made viewable on the handheld's screen. What's more, the handheld updates the tag's information, to indicate that it was visited by this work crew at this specific time. The handheld collects from the workers other information about the tasks they have accomplished and, when the workers are done, shows them the next stop on their appointed rounds. The mobile computer can collect information about any spare parts that are needed, and it can indicate that a particular repair may need to be rescheduled. When the crew gets back to their home base, the handheld is again synched with the main maintenance system and all the information it has collected during the day is transferred.

The key element of this maintenance management system is its use of a mobile application—software, or intelligence, that can be taken directly to the job site and, through RFID, gain a certain awareness of that site. This use of a distributed application, with logic located directly in the hands of repair people out in the field, has several benefits: It takes a good deal of pressure off the central system, which in theory could be handling these same tasks via a wireless link to a handheld "dumb" terminal. In addition, the self-contained handheld can operate successfully in the relatively harsh environment of the airport.

The Fraport maintenance system has been a great success, as can be seen when it is compared to the process it replaced. Previously, workers had to fill out four-page reports on the work that had been done and any problems still outstanding. At day's end, they would turn in their maintenance sheets to a central office where personnel, as quickly as they could, deciphered their handwriting and entered the reports into a computer system. Annually, the maintenance workers generated nearly 100,000 pages of such information, which, by law, had to be archived for the next 10 years.

The inevitable results of any such paper-based process were errors, uncertainty, delays, and unnecessary costs. Errors occurred in the paper orders and plans given to the maintenance worker, in identifying fire shutters or defects, and in data entry. This situation created unnecessary delays in carrying out repairs that were called for by safety inspections. And, in the end, there was no confirmation that the required maintenance work had been carried out.

CONCLUSION

Companies can leverage Real World Awareness to achieve a wide range of improvements in their businesses, from the incremental to the truly innovative and even revolutionary. For some, RFID will merely offer a way to better monitor shipments leaving a factory or to better track pallets within a warehouse. For other companies, RFID will provide the foundation for a thorough rethinking of such vital processes as manufacturing itself or the sub-daily replenishment of store shelves. None of these improvements will become available overnight, however. They will require effective planning and a strong understanding of the technologies involved. What's critical to remember is that each step forward on the road to Real World Awareness will bring with it substantial risks and substantial value, which must be carefully weighed against each other and kept in balance.

5

Implementing Real World Awareness

Transforming your company through Real World Awareness can take many forms, from a few incremental improvements to a wholesale restructuring of key portions of your operations. Simple changes are easily accomplished and digested, but, as the size of a project grows, so do the risks that are the focus of this chapter. Face it: For good reasons, the larger the project, the more everyone in the company worries.

After all, the excesses of the dot.com boom have left a bad taste in the mouths of executives of most companies as they look back on certain ill-advised IT investments that were made in a hurry and with the wrong focus. For a couple of years since then, economic hard times have made cost cutting the focus of IT. Some naysayers have even recommended giving up on IT altogether as a means to create competitive advantage.

On the other hand, the aroma of profits coming from companies like Dell, Wal-Mart, Purdue Pharmaceuticals, and other market leaders is starting to increase the appetite of senior management for IT innovation. All these companies have major investments in Real World Awareness technology that are either already in place or expected to improve operations that are already widely admired.

In addition, a steady stream of mandates is starting to appear from customers and channel masters who are eager for increased efficiency and visibility in the supply chain. The demand by the U.S. Department of Defense, METRO, and Wal-Mart for the increased use of RFID by their business partners is only the largest and most publicized of a growing number of Real World Awareness initiatives being promoted by key players. Government agencies, like the U.S. Food and Drug Administration, are creating regulations about such matters as the pedigree of drugs that will most likely be implemented by using some sort of Real World Awareness technology.

The question facing a growing number of business and IT executives is "How can projects, both large and small, to implement Real World Awareness solutions be carried out successfully with a minimum of risk?" Executives have realized that the transformation of business by Real World Awareness is not merely a single project or a short-term concern, but will be a continual source of pressure to innovate and re-create business models. The difficulty of implementing Real World Awareness is amplified by the multifaceted challenges of creating both new business processes and new partner relationships that use new architectures, technologies, and methods.

It is beyond the scope of this chapter to present a manual of tactics that covers every single challenge and technology problem related to implementing Real World Awareness. Rather, this chapter provides advice and warnings about potential problems that can help you succeed in adopting new and emerging technologies, like Real World Awareness. This chapter contains the essentials of the experience that a multitude of companies have undergone in deploying advanced technologies. It also presents the insight of the SAP support-and-consulting organization, one of the most experienced advisors regarding the adoption of new technologies. Although the details of your implementation will be unique, the habits of thinking that lead to success are similar from project to project.

This chapter might seem scary because it identifies a multitude of problems and provides numerous warnings, most of them aimed at getting larger projects done right. Although this information might have the effect of turning some people away from Real World Awareness projects, the fact is that most of us have no choice. We must become competent in this new world in order to compete and work with our partners. As with every other wave of new technology, what seems scary now will be well understood the next year and then easy the following year. And, as always, those companies whose employees learn and adapt quickly will have the advantage.

To tell this story, this chapter's first section looks at common patterns of failure for new technology projects; the second section examines the forces at play in Real World Awareness projects and summarizes some lessons learned from early adopters. The chapter concludes with a summary of advice distilled from early adopters and SAP consultants about how to keep complicated projects involving new technology on track.

THE TYPICAL FAILURE OF NEW TECHNOLOGY PROJECTS

For most companies, the mere thought of a large IT project already puts everyone on high alert. In many cases, implementations of innovative software have failed and then sparked a return to an extremely conservative approach to new technology. Although this situation might be right for many areas of a company's business, the ability to be a front runner in technology adoption in selected areas is a key competitive advantage. Stepping away from innovation means lost opportunities that can limit a company's ability to succeed. The damage done by a failed innovation project goes way beyond its direct impact: It can prevent a company from making the right choices for a long time.

With Real World Awareness, the risks are greater than normal and the pressures and complexity of implementations are fertile ground for mistakes of all kinds. Some are caused by the fact that Real World Awareness is a new technology. The failure of new technology projects follows a pattern caused by the same set of forces that lead to the same mistakes in every generation of new technology. Other problems are unique to Real World Awareness technologies.

To provide the context for this discussion of how to avoid disaster, this section describes a few negative patterns that show how standard operating procedure can mean death to Real World Awareness projects.

Part of the problem is the high level of excitement about Real World Awareness. Everyone involved in an implementation, especially engineers, becomes excited about its potential, they "fall in love," and many of these events start to happen:

- The scope grows out of control because everybody wants to use the new technology to solve problems.
- Every project attempts to become related somehow to the new technology, to take advantage of budget allocations and employee enthusiasm.
- Business goals are forgotten or glossed over, and the technology takes center stage.
- Strategy becomes defined by the technology.
- Real experience is scarce, yet many vendors jump on the bandwagon and claim competence.
- Technology is implemented in a quick-and-dirty fashion, without proper consideration of a long-term architecture.
- Software development projects are rushed into production without proper quality assurance.
- Because projects become highly visible, the cost of failure is high, which leads people to hope for a miracle, even in the face of serious problems.
- The difficulty of changing an organization is underestimated and not adequately planned for.

All these pressures can add up to an assortment of failures. This section presents three typical patterns.

THE SCOPE SPINS OUT OF CONTROL

An aggressive young project manager is put in charge of the implementation of new technology. He focuses too much on the glory of success and not enough on the threats to the project. At every meeting in the early phases of the project, the project manager is happy to extend the scope of the project because that will lead to an even larger victory. Engineers add fuel to the fire as they try to introduce requirements that will give them experience with every aspect of the new technology. Advanced and unproven technology is added at several places. The result of these decisions is invariably a late project, chock full of integration problems that lead to endless deadline extensions and, in many cases, complete failure.

QUICK-AND-DIRTY ARCHITECTURE

A new technology is forced on a company by a large customer. The systems integrator, who was chosen on a fixed price and who is using the project to gain experience, recommends a custom integration around a niche vendor. The initial implementation succeeds, but a year later the company must expand the scope of the new technology. This process requires expanding the integration to several key business systems. Accomplishing this expansion with the current architecture means a huge integration bill and a large total cost of ownership (TCO) to maintain the custom code. The entire first implementation is ripped out and replaced with technology that is much easier to integrate with the business systems.

CHANGE-MANAGEMENT OBSTACLES

A new warehouse-management system is implemented with RFID technology to speed inventory tracking. After the system is launched, the warehouses descend into chaos as the technology for reading inventory fails. At first, the readers appear to be unreliable, but further analysis shows that they are being sabotaged by workers fearful of losing their jobs. A gradual 10 percent reduction in force was planned, but management expected to achieve it over the course of a year through natural attrition. The system works finally, three months after its intended launch, after an extended communication and training program overcomes initial hostility to the technology.

The mistakes in these examples will ring true to most veterans of large IT projects. The challenge lies in maintaining a constant awareness of all threats, which can be difficult with Real World Awareness because the threats are so numerous.

MANAGING THE RISKS OF REAL WORLD AWARENESS PROJECTS

Implementing Real World Awareness solutions involves many forces that turn an already daunting process into an extreme challenge. In almost every Real World Awareness project, these factors loom large:

- **Mission-critical systems.** Real World Awareness implementations are usually focused on the key systems used to create value or to support important customers. The stakes in making changes to these systems are the highest of any sort of IT project and have the greatest risks and rewards.

- **Multiple disciplines.** Real World Awareness projects involve almost every technical skill in a company. Electrical and mechanical engineering used to create tags must be coordinated with the placement of radio-based readers that interact with manufacturing lines. All this is controlled and monitored by software. Keeping the big picture in mind is a huge challenge. If no one understands the big picture, contradictions and inconsistencies will be found late in the implementation process.
- **New processes.** Real World Awareness solutions almost always introduce new types of processes in an organization. Executives and technologists must learn to manage processes that cross company boundaries and harvest information from systems distributed all over the world.
- **New partners.** Real World Awareness solutions are usually implemented with a set of contributing companies that are working with each other for the first time. All communication, project management, and working styles, therefore, must be learned and understood, and new contractual arrangements must be created. Both factors can cause delays.
- **New architecture.** Real World Awareness solutions frequently represent the most advanced and comprehensive application of web services in a company. Although web services have gradually appeared in IT environments, operating them at a scale to precise service levels usually involves a learning curve, during which expertise is established.
- **New technology.** Real World Awareness solutions almost always introduce new hardware for tagging items and harvesting information, and they frequently greatly expand the production use of wireless networks.
- **New lines of business.** Real World Awareness frequently opens up the opportunity for new services or entire new businesses.

In Real World Awareness implementations, almost all these factors are in play at one time, which means that projects for implementing Real World Awareness generally start out in a high-risk state. Add to that the high expectations for a new technology or the demands of a mandate, and a project can easily become a pressure cooker.

Here's the spot where you might expect to find the secret to successfully implementing Real World Awareness projects. No such secret exists, although experience is being gained every year that one day lead to well-understood best practices. Until then, the forces mentioned in the preceding list are conspiring to create some remarkable findings in Real World Awareness implementation. Consider the warnings in this section, which were gathered from early adopters in 2004.

Tag Availability

After an RFID tag or other device becomes attached to a product, the tag becomes part of the bill of materials for that product. If you don't have tags, therefore, you cannot produce the product. Early adopters have found that demand is rising at such a rate that getting tags on time can sometimes be challenging.

Tag Quality

The manufacturing process for RFID tags is complex, so a certain rate of errors is to be expected. Early adopters in 2004 reported that this rate could sometimes be as high as 10 percent, which means that the processes for including tags in manufacturing must have a step for quality control and inspection.

Read Tags Early and Often

After tags are placed on items being manufactured, many different steps can follow. To ensure that the tags aren't damaged during these steps, it is wise to design processes to read the tags several times as a product is being created. One drug maker that tags its bottles with RFID tags reads them at least five times before the final read event, at the end of the assembly line, and notifies the manufacturing system that a new product with a new ID number exists.

Disrupt Core Processes As Little As Possible

Most manufacturing processes have been finely tuned and optimized to produce as fast as possible. The introduction of Real World Awareness should try to preserve as much of this progress as possible, which might require ingenuity. One pharmaceutical company chose to add RFID tags to its labeling process rather than interfere in any way with a manufacturing process that loaded 120 bottles per minute.

READING ACCURATELY IS AN ART

Here is what won't happen during RFID implementation. You set up an RFID reader and pass RFID tags by it, and everything is always read correctly. Instead, a variety of factors impede the technology. All sorts of surfaces turn out to reflect the radio beams sent out by readers. Machinery and other devices interfere with proper reading. The tags must pass by the readers within the right distance and at the proper speed. At the time this book was written, in 2004, configuring readers and making sure that the tags pass by within range was a rapidly progressing art that is slowly progressing to a predictable science.

ELECTROMAGNETIC SPECTRUM TRAFFIC JAM

The electromagnetic spectrum is more crowded than you might think. One of the first things engineers learn about RFID tags that operate at 915 MHz is how close that number is to the range at which Wi-Fi and WLAN wireless networks operate. RFID and wireless technologies can easily interfere with each other, and untold numbers of devices involving whirling metal can produce radio waves at many frequencies. Debugging these sorts of problems requires many different skills and is frequently difficult.

READER ANOMALIES

Even if readers are performing perfectly and tags are passing by them in lock step, many other strange things can happen. For example, a pallet left too close to a reader might be read all day long. At one plant, strange readings resulted when an employee walked by readers with spare tags in his pocket.

TRAFFIC PATTERNS MATTER

To prevent strange behavior, warehouses and other locations using Real World Awareness technology must be stricter about where equipment carrying tagged items can move. A forklift carrying a pallet of tags can no longer drive by a line of readers at a loading dock without causing problems because the pallet is reread by each one.

DEVICE CHOREOGRAPHY

Leaving readers on all the time is frequently a bad idea. Some regulations prohibit workers from being exposed for too long to radio waves. Early adopters have found that creating a sequence in which Real World Awareness devices are turned on and off helps in many cases to increase efficiency and address safety concerns.

MODELS GO WHERE NONE HAS GONE BEFORE

One effect of the introduction of Real World Awareness in any environment is an increased appreciation of the complexity of all the different layers and how they interact. The physical layout of a plant must be coordinated with the electromagnetic spectrum, as well as with other machinery. All these parts are generally represented in a virtual model in a software application. One early realization by engineers in a Real World Awareness implementation is that current models of the plant must be supplemented. Creating these models in software can be a significant challenge that should not be done in a hurry.

ANTICIPATE MANDATES AND DOWNSTREAM REQUIREMENTS

The availability of information from using Real World Awareness changes things. After RFID tags are available on a product, they tend to get used. Early adopters suggest that the introduction of Real World Awareness anticipates the natural demands of the expected next steps so that they can be accommodated easily. Designing to satisfy a narrow mandate is usually a mistake.

These are just some of the interesting findings that have been collected from the field. Each month, more is being learned.

SUCCEEDING WITH REAL WORLD AWARENESS PROJECTS

The different strengths and weaknesses in business processes, different competitive advantages, different business partners, and different hardware and software in use at companies make a general one-size-fits-all road map to Real World Awareness implausible across industries or even across divisions within a company. Although no silver bullet exists, there are some general patterns to success.

Although all the advice in this section is sound, some of it might not be new to you. On the other hand, many of these recommendations are known but not followed in many projects, so they bear repeating.

Focus on Process

One of the most important best practices is a process-centric approach. The ERP and supply chain implementations of the past 15 years have taught us that adopting technology-driven concepts like Real World Awareness without identifying and implementing improvements in business processes can harm a company's performance far beyond any money wasted on IT. For Real World Awareness projects, a process focus can be challenging because the processes being automated are newly designed or cross company boundaries, or both. The process must be understood and communicated thoroughly.

Beware Technology Strategies: Begin and End with Business Problems

A company should shy away from identifying a business strategy with a technology label. An RFID strategy should be thought of as a contradiction in terms or an incomplete statement, one that is missing the point that technology exists only to meet business goals. A keen focus on how a business should evolve must be primary. The technology must remain secondary, even if the new business process is made possible only by new technology. With exciting Real World Awareness technologies, such as RFID and sensor technology, you can easily get carried away by tactics and lose your strategic vision.

Turn a Vision into a Road Map

The tremendous value that Real World Awareness techniques like RFID can achieve makes it acceptable to be excited about technology again. This excitement is a positive force that should be cultivated, but it easily could become a distraction from business problems.

Colgate is a $9.9 billion consumer products company, based in New York, that operates in more than 200 countries. Approximately 70 percent of sales come from international operations. As a manufacturer of hundreds of different products that must distribute products to major retailers around the world, Colgate has a multitude of Real World Awareness challenges in running its operations.

Ed Toben

Chief Information Officer

Colgate

Ed Toben is the CIO of Colgate, which has been a leader in the implementation of emerging technology. As a company, Colgate has repeatedly succeeded where many have failed in the rapid implementation of solutions early, before the products were at full maturity and before best practices were known to the general marketplace. Ed received his BS degree in business administration from Villanova University and his MBA degree in finance from Fordham University.

Q&A with Ed Toben, Chief Information Officer, Colgate

Q: What is your approach to implementing emerging technologies?

A: It's really about focus and execution. It is very easy to lose your focus because technologies and projects are very large and very complex. It sounds simple, but is hard to achieve in detail. Our first step with SAP was to implement what was then the new R/3 in Colgate's U.S. business. Our business need was really clear, so we succeeded. For Real World Awareness, we need to think of our supply chain as an integrated process and make the business need crystal clear. We are careful in all of this to keep our eye on the business goals so we can make sure we are not distracted by how cool all this technology is.

Q: What business goals are you trying to achieve with Real World Awareness solutions?

A: We are taking the time and money out of our order-to-cash process and improving the service to our customers. Retail Partners demand an increase in the speed and accuracy of the supply chain. We want to keep improving our growth margin, so we can invest back in our own products. If you take Real World Awareness and RFID, the way we look at it, it's just the next step. We started with R/3 integrating the base and then optimizing our processes and performance with SAP Advanced Planning and Optimization (APO). Real World Awareness, such as RFID, is the logical extension to all that.

Q: What have you learned during implementation?

A: The secret is really picking the right partnership, and in our case it's with SAP. You must know how your partner is looking to the future. What is their vision? Is it aligned with your needs? What are they spending on R&D? At Colgate, our core competency is developing new consumer products, not new technology products. We make a conscious effort to stay close with the partners who do that and provide the whole solution, including integration. Why should Colgate spend its time trying to make the applications work together? That's what we want SAP to do, so that we can put our efforts into implementing all these things.

Q: How do you maintain focus and keep control of large projects?

A: Projects involving new technology are always bigger and more complex than they appear. We control scope by keeping delivery dates firm and then reducing scope to make them. The business impact of late delivery is much larger than most people realize.

Q: How do you manage the risks of large projects involving new technology?

A: We try to get our partners involved and stay aligned with them. We seek to design our roadmap to maximize the core direction of our vendors' products. We don't want to be the only customer for anything. It is also really important to communicate our strategy and goals. We have 900 people in IT, in about 60 countries.

A business strategy that employs Real World Awareness technologies must be clearly defined, thoroughly thought through, and communicated properly. A long-term vision should lead to a three- to five-year road map and, finally, to a detailed plan for the following year. Over-communicate about the whole plan, and focus on your business goals. Don't overestimate your organization's ability to digest change. Then, execute, execute, execute.

INCREMENTALISM: THINK BIG, START SMALL

The pace of mandates and the arrival of new government regulations have caught many companies unprepared. It is not uncommon, therefore, for Real World Awareness projects to be implemented in a rush. Although this is true for many IT projects, the challenge with Real World Awareness is more acute. The problem is that Real World Awareness projects tend to cross many boundaries: between IT and manufacturing, between departments in a company, and between companies cooperating in a network. If saving money and building to meet future needs is a goal, some level of comprehensive architecture must be designed before even the smallest project is pursued. Thinking big now avoids having to start over later. Starting small means that requirements and assumptions are tested with the smallest possible investments and skills are honed to reduce risk on future projects.

BUILD A LONG-TERM ARCHITECTURE

New technology is often implemented in a hurry. Ignoring architecture or rushing to implement the cheapest quick-and-dirty solution frequently leads to a situation in which a system is completely replaced soon afterward, as integration and maintenance costs mount in the face of new requirements. If you "buy cheap" or design poorly, you often buy twice.

Create an architecture to accommodate change. Speed and flexibility must be planned for. Vet the architecture by asking how future change will be accommodated. Make sure that vendor road maps lead to a more flexible system.

SEEK TRANSPARENCY AT ALL LEVELS

Transparency means, at its core, that an accurate model of a project is there for all to see. This model becomes the basis for fact-based decision-making and the early identification of problems and leaves less room for assumptions, mistrust, and politics. But management must want to see and accept the truth.

There is always a lack of transparency in failing projects. Scope is a mystery, as is the project status and an understanding of who is responsible for what. When deadlines are missed, nobody knows why.

REALIZE THAT HOPE CAN BE THE ENEMY

In the face of disaster, hoping for a miracle provides a wonderful way to avoid the unpleasant truth about what is happening. When projects start to slip, hope steps in. Critical issues are identified, but left to fester. Employees might be afraid to tell their bosses or internal rivals the bad news. At some point, usually near a deadline, reality must be faced and the ugly mess is visible to all.

Management must make clear that bad news is acceptable. Project steering committees must be designed to search for problems, identify risks, and then take decisive action.

DEFINE CLEAR GOALS

Don't lose sight of the purpose—it is not technology. When a project gets in trouble, it is easy to forget the goals and just abandon the whole thing.

The primary goal must be to achieve a project's business objectives. When things go wrong, however, what then? If projects get into critical situations, some people walk away from the risk of failure. The second goal must be to achieve as much as possible in the face of large problems that prevent complete success. Making half of a project happen is better than writing off the entire thing. Be crystal clear about your motivation. Share an accurate picture of what has happened with all relevant stakeholders, and try to make a collective decision about what to salvage.

CONTROL SCOPE TIGHTLY

Be realistic. Use the 80/20 rule, in which 80 percent of the value comes from 20 percent of the features. Be careful, though: 80 percent of pie in the sky is still pie in the sky. Projects dealing with new technologies are a magnet for excitement, scope creep, and unreasonable expectations.

Simplify and reduce scope at every possible moment. Even successful projects can be considered a failure when they are measured against implicit, unrealistic expectations. Pilot smart: Choose pilots that have definable scope, and then use the results to carefully scope larger projects. Communicate the scope widely.

DON'T MOVE DEADLINES—REDUCE SCOPE

Projects with great potential always get bigger and more complex than originally anticipated. In many cases, the most enthusiastic people on a project drive this expansion as their ambition to make the project a huge success gets out of control. It is tempting then to move the launch dates until the grand dream is accomplished.

In the face of a looming deadline, reduce scope to make the launch date. Postpone features to Phase Two. Communicate the change in scope. Getting a project completed and in production gives a company concrete benefits from having a working system as well as a wealth of information about how to get it right.

STREAMLINE DECISION-MAKING

Steering committees populated with a large number of people have to reconcile a large number of interests. This situation frequently makes cutting scope impossible because one person doesn't get what he wants. Clear priorities are hard to set and decision-making slows to a crawl because each issue becomes a political dogfight.

Steering committees must be created from business, IT, and practitioners on a project, empowered explicitly by top management to make far-reaching decisions without delay. Keep this group as small as possible. Face political issues directly. Set goals for the project, and let the steering committee move quickly toward meeting them.

CREATE STRONG PARTNER RELATIONSHIPS

When vendors are worried about sustaining and nurturing relationships with clients, they are fearful of delivering bad news. Unless vendors are directed to be openly critical, they will hope for the best. When a vendor is told to do a narrow job and not think about the larger success of the project, it doesn't help find and manage risks outside this domain.

Choose vendors who are seeking, and have established, long-term relationships with their customers. Bring them into the team and encourage them to seek out problems. Be a demanding but rewarding customer.

REDUCE DEVELOPMENT DEPENDENCIES

The development of new custom software is frequently required, especially with new technologies. The cost of developing software, compared to the budget for many IT projects, is relatively low, but the length of time it takes to complete can be unpredictable. Many large projects are delayed because software development takes longer than expected, which can be costly.

Limit dependencies, if possible. The biggest possible buffer should be planned for. Limit features in the initial release while planning for the next version. Keep in mind that software developers are optimists by nature and add a little more cushion to the schedule than you think you might need.

PLAN CAREFULLY FOR THE INTEGRATION TEST PHASE

When all the parts of an RFID project are first brought together, many issues appear for the first time. Frequently, at this point, key technical resources might have moved to new projects. Users are not patient, and exposing them to "broken" software might make RFID adoption more difficult.

Plan ahead for difficulties in the integration test phase. Make sure that those with the necessary know-how about the project remain available during this difficult phase. Search for professional help—consultants can provide experts in this phase who can dramatically reduce the risk of failure.

Keeping on the lookout for these risks can help you become much more successful in leveraging Real World Awareness and other new technologies to achieve business goals. The bad news is that just knowing the rules does not solve the problem. Success depends largely on your ability to execute your plans and decisively take the steps needed to adapt to a changing environment. Although the effort to implement Real World Awareness involves significant effort and careful attention to the risks involved, the rewards are huge. Companies in almost every industry are, despite the challenges, repeatedly proving that success is within reach.

6

People, Privacy, Politics

Whereas previous chapters outline the pitfalls of implementing Real World Awareness technologies and how best to avoid them, this chapter discusses the larger social context in which these technologies will function. What is the big-picture value of Real World Awareness for industry in the supply chain and beyond? What consumer benefits are waiting to be unlocked in areas like security, health, lifestyle, and convenience? And, what obstacles stand in the way of large-scale adoption and long-term transformation?

A recognized piece of IT wisdom about the way society perceives imminent technological change says that people tend to overestimate the disruptive impact of new technologies in the short term, while underestimating their far-reaching impact in the long run.

Whoever said it first certainly had the Internet in mind: The assertions of the boom and the subsequent bust preceded the long period of emergence we are in now. Real World Awareness technologies are on a similar path, with one notable exception: In this instance, the technologies' champions are the ones making modest claims for the near future; Real World Awareness, they vow, will optimize supply chains first and then change the world later—assuming, of course, that the cost of tags falls to around five cents each, which may happen by 2012, or maybe 2008, or even sooner.

Critics of Real World Awareness, on the other hand, seem convinced that its impact will be immediate, pervasive, and powerful—and that this is a worst-case scenario. Indeed, the emergence of Real World

Awareness has alarmed civil liberties and privacy advocacy groups, like the American Civil Liberties Union (ACLU), Consumers Against Supermarket Privacy Invasion and Numbering (CASPIAN), and their allies around the globe. Several groups have already organized boycotts against RFID pioneers, and now are working to introduce legislation that would mandate, for example, how RFID tags may be used by retailers.

The concerns of these groups, ironically, are based on extremely confident predictions of Real World Awareness' future capabilities: that nearly all RFID scanners will be able to read tags at distances of 30 feet or more; that scanners will be ubiquitous and inexpensive enough to be readily available to governments, corporations and private individuals; and that businesses will rush to integrate Real World Awareness with their core systems and with each other without considering the privacy implications and consumers' qualms. When all this is achieved, the privacy groups maintain the result will be privacy invasion, government surveillance, and consumer exploitation on a massive scale.

It's no wonder, then, that the ACLU and CASPIAN have even called for a total moratorium on item-level RFID tagging in the United States until Congress can pass laws explicitly protecting citizens from intrusion by Real World Awareness technologies. If this scenario were to happen in the short term, the disincentives to utilize Real World Awareness technologies would likely cripple their potential before the long-term benefits even began to take shape. It's not hard to see why some technologists are frustrated by RFID opponents' misconceptions and by the current lack of middle ground in the ongoing privacy debate. Studies have already shown that consumers' attitudes toward Real World Awareness are linked to how knowledgeable they are about the technical workings of RFID and similar technologies. Well-informed consumers are less prone to dwell on the negative possibilities and more likely to realize the benefits that come with any privacy tradeoffs.

Although IT managers are used to fighting for the adoption of new technologies, they must appreciate that this battle goes beyond making a business case for installing RFID readers in a warehouse. Unlike the vast majority of projects, Real World Awareness technologies are potentially so transformative in the long run, that they come with a set of societal and political issues on top of traditional concerns like technology adoption, integration costs, and ROI.

With every new innovation, the benefits also come with a down-side. The Internet has created a global public network, and with it came viruses, spam, unmonitored access to pornography, and other negative aspects. Credit cards brought us all security and convenience, as well as fraud and digital records of spending that can be abused in the wrong hands. Mobile phones connect us anywhere but also threaten the tranquility of performances and other public spaces. Thus it will be with Real World Awareness, and the more consciously we come to grips with both the benefits and the drawbacks, the better for all involved parties.

This chapter examines some of the conflicting projections about RFID tag costs and availability—the widespread use of a tag that costs five cents or less is believed to be the "tipping point" for Real World Awareness—and pays particular attention to the raging privacy debate.

Although privacy advocates call for an outright moratorium, some states are considering legislation, and Real World Awareness proponents argue that their concerns are overblown, fed by mythical assumptions about RFID's powers. Meanwhile, RFID standards bodies, like EPC-global, and technology providers are proposing their own principles and standards for Real World Awareness technologies. It is their hope that self-regulation, combined with better consumer and industry education, will ultimately reassure anxious users.

This chapter is divided into four sections. The first touches on the value that Real World Awareness has already or is about to unlock for industry and for consumers. This section argues that the key to winning consumer confidence is to demonstrate that the benefits of Real World Awareness far outweigh the inherent issues and how a phase of consumer concern and uncertainty is traditional in a new technology's life cycle. (Consider early fears about the security of e-commerce or the surveillance of analog cell phones.)

The second section delves deeper into privacy and policy concerns by outlining the positions of Real World Awareness critics and support-ers, proposed legislation, and the need for consensus about what an empowered and informed consumer's rights are and what can be expected from the creators of Real World Awareness technologies.

The third section takes a look at potential barriers that could hin-der the widespread adoption of Real World technologies, such as tag prices and looming standards battles.

The chapter concludes with an examination of what we can expect from the intermediate- to long-term adoption of Real World Awareness. What will be the pace of change? What principles of development and use should adopters of Real World Awareness follow?

The ultimate goal, of course, is to strike a balance between respecting consumers' fears while continuing to drive development and adoption of Real World Awareness technologies. While consumers might be concerned with the possibility of privacy violations for now, in the long run they will no doubt welcome lower prices and supply chain efficiencies resulting from widespread adoption. Two things in particular will help make this happen:

- An ongoing dialogue between technology providers, corporation users, and consumers about privacy guidelines and best practices for using RFID and similar technologies. Trusted third parties—standards bodies, and industry associations, for example—must bring all parties together for transparent discussions about what's at stake.
- Continued efforts to educate consumers about both the technological limitations of Real World Awareness and the benefits stemming from these technologies as they exist now. Helping consumers become familiar with Real World Awareness technology will ease the transitional period and help them feel more comfortable with the concept later on.

PROVING THE VALUE OF REAL WORLD AWARENESS

This section examines the benefits of Real World Awareness technology for industry, including supply chain improvements and the often unexplored benefits that this technology offers consumers in the areas of security, health, and convenience.

THE VALUE FOR INDUSTRY

Making the case for the value of Real World Awareness to retailers, manufacturers, and even IT suppliers (who are poised to profit handsomely from its adoption) is becoming easier by the day.

The mandates of retail chains such as Wal-Mart, METRO Group, Target, and Tesco that their suppliers meet pallet- and case-level tagging requirements over the next few years have legitimized the potential benefits of Real World Awareness for supply chains and other internal business processes in the eyes of many IT professionals. Additional mandates and initiatives by the U.S. Department of Defense to its contractors, Airbus and Boeing to their suppliers, and the Food and Drug Administration to pharmaceutical makers have further galvanized interest.

Although retailers have taken the lead in promoting widespread adoption, even manufacturers and retail chains that are not linked to any of these organizations have begun exploring how to optimize their own supply chains and internal business processes. Although even these companies are moving cautiously and are still uncertain about the ultimate ROI of current-generation implementations, pilot projects—like METRO Group's Future Store Initiative and other, less heralded ones—have highlighted potential benefits for businesses in these areas:

- **Inventory control, distribution, and order fulfillment.** With more accurate, real-time data at the pallet and case level, manufacturers and retailers can track and verify shipments more efficiently and smoothly and eliminate the waste of missing, stolen, and incorrectly assigned goods. Automation of the auditing process also reduces labor costs. One analyst has predicted that if the use of RFID becomes widespread, retailers could reduce standing inventories by 5 percent, warehouse labor by 7.5 percent, and product losses by 1 percent of sales.[1] Considering that retail trade in the United States totaled $1.09 trillion in 2002, those savings are significant.
- **Asset management.** Tagged assets can self-monitor their usage, by providing data that can be used for the more effective maintenance and replacement of those assets, thus lowering costs. The RFID technology provider and asset-management company TrenStar owns 65 percent of the beer kegs in the United Kingdom and used its RFID network to accurately model the best time to service a keg tap, which led to a fivefold decrease in maintenance costs.[2]
- **Manufacturing optimization.** RFID-tagged assets can also help with process optimization during the manufacturing process, as opposed to simply sorting the results afterward. At the IBM fabrication plant in Fishkill, New York, RFID-tagged component containers are tracked and routed by a central network to optimize chip production and reduce staff by 50 percent. Infineon, the European chipmaker uses RFID to track work in progress. At the other end of the spectrum, the French pharmaceutical/medical devices company, Dentalab, now sells an RFID-enabled system to dental laboratories for the manufacture of dental crowns. Molds have an RFID tag implanted in them to guarantee the identities of their intended owners, thus minimizing the number of missorted crowns.

- **Authenticity and anticounterfeiting.** The Food and Drug Administration (FDA) mandate for pharmaceutical manufacturers to use RFID-tagged pill bottles was driven by the desire to eliminate the estimated 2 to 7 percent (approximately $30 billion) of counterfeit drugs sold each year. The tags will allow for the tracking and verification of a drug's origins and journey to the shelf. RFID tags will also be implanted in tickets for the 2006 World Cup football tournament, in an effort to thwart counterfeit tickets and speed fans' entry into the stadiums.
- **Individualized sales and marketing.** Some of the more tantalizing ideas sparked by METRO Group's pioneering Future Store Initiative include *smart shelves,* which not only track inventory levels, by sending a notice to store staff whenever they need to be restocked, but also trigger a smart advertising display nearby. Real World Awareness technologies could potentially transform the passive data on shopping habits collected from store club cards into real-time, actionable data that can be used to offer special deals, for example. Mapping GPS data to individual cell phone users could enable Barnes & Noble, for example, to offer a limited-time (maybe 20 minutes) 30 percent discount to a previous customer walking near a store. But these scenarios are technically still far away because they require a degree of data and application integration currently lacking a proven business case.
- **Aftermarket services and product recalls.** Sending consumers home with live tags with continued integration into retailers' systems offers myriad benefits in the long run. Live tags could be used to store warranty information or directions for care and assembly. Or, the tags could be used to single out specific items as part of a customer recall—individually tagged items brought back to manufacturers by concerned consumers could be easily cross-checked with the recalled product series. In some areas, livestock is already being implanted with RFID chips that store data about a given steer's owner and origin. That data could prove instrumental in tracing the path of an outbreak of mad cow disease, for example.

THE VALUE FOR CONSUMERS

Although the value of Real World Awareness for industry is an increasingly hot topic at IT conferences, in trade journals, and within companies, an articulation of the benefits to consumers definitely lags behind. No one has yet offered a coherent, comprehensive outreach effort to consumers that stresses the benefits of these emerging technologies, even though consumer acceptance is obviously critical. This section looks at some of the benefits of the sort identified in Figure 6.2.

Value Chain Area	Business Process	Value Driver
Service Management	Asset Maintenance	Time and Costs, Safety
	Spare Parts Inventory Management	Customer Service Levels, Inventory Turns and Days-on-Hand
	Warranty Management	Fraud Costs
	Returns Management	Transportation Costs, Inventory Write-offs
	Repair Management	Customer Service Levels, Time and Costs, Safety
Sales and Marketing	Collaborative Demand Planning	Forecast accuracy, out of stocks
	Trade Promotion Management	Retail Execution, Compliance
	Shelf Replenishment	Out-of-Stock
	Value-added Services, Like Smart Scales, Self Checkout, and E-Coupons	Customer Service
Distribution	Picking and Packing	Time, Accuracy
	Shipping	Time, Accuracy
	Replenishment	Customer Service Levels, Inventory Turns and Days-on-Hand
Manufacturing/Sourcing	Assembly Control	Production Time, Labor, Quality
	Replenishment	Inventory Levels, Lead Time

Figure 6.1. RFID Business Process Opportunities and Related Value Drivers

RFID Application Areas	Consumer Value
➤ Food and Drug Safety ➤ Health Card ➤ Body Value Measurement	Better Health
➤ Pet Tagging ➤ Car Condition and Location ➤ Aircraft Maintenance	Better Security
➤ Smart Shopping Cars and Self Checkout ➤ Customer Shopping Card ➤ Store Replenishment	More Time and Money
➤ One-to-One Marketing ➤ RFID-based Ticketing ➤ RFID-Enabled Cell Phones for Local Information Access	Improved Convenience and Lifestyle

Figure 6.2. RFID Applications and Consumer Value

SAP AG, the leading maker of software applications for business, hopes that its own active role in encouraging dialogue between technology providers, manufacturers and retailers, and consumers is one of a trusted third party with the credibility and technological leadership necessary to bring all involved stakeholders together on a global scale. SAP has already hosted open public discussions on the topic in Germany, in Brussels, and in Washington, D.C., and the company seeks to play a leading role in helping develop clear guidelines for the responsible and judicious use of Real World Awareness technologies in the future. SAP is also playing a role in encouraging the broad adoption of Real World Awareness by educating consumers about the technologies' myriad benefits. Although sketching the outline of an educational campaign is beyond the scope of this chapter, consider some of the following examples of real consumer benefits in the areas of security, health, and convenience that can and should be used in making the case for Real World Awareness.

Security

In the spring of 2005, the United States State Department will begin issuing "smart passports" to U.S. citizens. These smart passports contain RFID chips that store a passport holder's name, date of birth, and place of birth. The chip, which holds 64K of memory, also has enough room to store biometric data, including digital fingerprints, photos, and iris scans. The technology, which may also be adapted for use in driver's licenses, is part of the United States government's efforts to secure its borders and fight terrorism in the wake of 9/11.

In Mexico, a number of government officials have had RFID chips implanted subdermally. The chips control access to areas containing sensitive documents and rather than entrust access privilege to identity cards that might be lost or stolen, these implanted chips have proven to be much more difficult to duplicate, and with improvements, may prove to be as reliable as retinal scans and similar measures.

A pair of elementary schools in Japan have asked parents to embed RFID tags in their children's bags, which are routinely scanned upon arrival at school each morning. After a handful of high-profile child murders and kidnappings shocked the country, this pilot project was conceived as an early-warning system for school officials, who felt that taking roll call took too long and happened too late in the day.

Commercial air carriers, such as Delta Air Lines, are turning to RFID and sensors as effective tools for aircraft maintenance. Tagged aircraft components could carry maintenance history data or even detect and report unusual performance via onboard sensors. Not only would airlines benefit from more cost-effective maintenance routines, but passengers would also receive the ultimate benefit of increased safety while ideally flying less expensively.

Health

Pending a full-year review, the Applied Digital VeriChip was approved by the FDA in October for implanting into humans. The RFID chip, about the size of a grain of rice, will be marketed to sufferers of Alzheimer's disease, diabetes, cardiovascular disease, and other conditions requiring specialized care. The chips have a unique identification number and use EPCglobal's Object Name Service to allow a server to be identified. The identification number can be used to locate on the server a data file that contains a patient's medical records and treatment instructions, for use in case its owner arrives at a hospital unconscious or otherwise unable to communicate his condition.

Future generations of RFID tags will regularly read and write data, tracing the lifespan and temperature of perishable goods. In the future, RFID-tagged pill bottles might warn a "smart" medicine cabinet that the pills inside have expired, and a smart refrigerator might notify its owner that the tagged carton of milk inside is two weeks old and may be unfit to drink.

As noted earlier in this chapter, the tagging of livestock (to help trace the spread of disease), the tagging of pill bottles (to fight counterfeiting), and the use of live tags in products (in the advent of a product recall) all carry health benefits for the consumers of these goods.

Experiments have also been conducted with RFID tags implanted in surgical instruments and sponges, to ensure that no equipment is inadvertently left inside patients at the completion of surgery.

Time and Money

In addition to health and safety benefits, consumers stand to gain from RFID in two very appealing areas: time and money.

- **Instant, invisible checkout**. Prefigured by the EZ Pass systems used on some toll roads in the United States, in which a long-distance RFID reader recognizes tags mounted on car windshields and charges their owners for the toll without the drivers having to slow down, the same concept can be applied to grocery stores, gas stations, or any retail environment in which an RFID reader can be linked to a customer's credit card number. Rather than head for a checkout lane, customers can simply head for the door via a scanner, confident that their purchases will be automatically billed to their cards.

- **One-to-one marketing**. Personal-recommendation software engines have become one of the most popular features of e-commerce sites like Amazon, which uses customers' purchasing histories to suggest similar products or product categories. RFID-enabled smart shelves could bring this idea to supermarkets or any other retail stores: A consumer's shopping card could be analyzed in real-time by store-based readers to generate recommendations for similar products or to point the consumer to promotions in the store, for example. The consumer would thus save time and money and the retailer provides to the consumer a value-added service: a more convenient shopping experience.

- **Lower prices across the board.** Another invisible benefit of smart shelves and similar Real World Awareness retail technologies are the savings realized by retailers, which are frequently passed along to consumers. As Wal-Mart has demonstrated for decades, savings realized by new technologies and process improvements quickly lead to lower prices. The savings realized by Real World Awareness will be no different, and consumers will reap the benefits.

- **Just-in-time reviews and recommendations.** The potential also exists for combining recommendation engines with real-world locations. Rather than drop by the home page of Amazon.com or look up the latest restaurant review on Citysearch, a tagged shopper might receive recommendations from an audio kiosk at Virgin Records, for example. The address and map of the hottest new restaurant in the neighborhood might be automatically delivered to a consumer's mobile phone.

- **A home network of smart appliances.** The smart refrigerator we mentioned in the preceding section is likely to become just one piece of a wirelessly networked smart household. RFID readers and other Real World Awareness components embedded in devices will be able to continuously monitor the state of food in the refrigerator, the washing instructions for a shirt (read by the washing machine off a tag embedded beneath a button), or even the whereabouts of the family cat (which has a subdermal tag implanted in case it gets lost).

STRIKING A BALANCE

Current resistance by privacy groups, legislators, and frightened consumers to the widespread adoption of Real World Awareness appears to have less to do with a fear of the specific technologies that comprise the concept than it does with historical resistance to every new development concerning the aggregation and linking of personal data. As time goes on, history has shown that resistance gradually gives way to wariness and then to acceptance and ultimately the feeling that these technologies had always existed.

Consider the Internet. Early efforts at building e-commerce businesses were stymied by many users' fears that credit card transactions were insecure and likely to be intercepted by hackers. This was despite the fact that even early Web browsers had built-in encryption schemes. After a number of early, failed experiments in creating alternative forms of Web currency, the rise of popular sites like Amazon and PayPal gradually habituated users to using their credit cards online. Consumers became comfortable with the decreasing risks.

And, before the rise of the both Internet and Real World Awareness technologies, protests were made against the spread of bar codes and even store club cards. The fact that interest groups failed to stop the adoption and acceptance of these technologies suggests that an opposing minority will always exist but that it is not an insurmountable obstacle.

For Real World Awareness proponents, it is certainly tempting to dismiss Real World Awareness critics as uninformed Luddites, but that simply isn't the case. The presence of the Electronic Frontier Foundation (EFF), a group founded by tech-savvy, pioneering cyberlibertarians, suggests a deeper anxiety that is specific to Real World Awareness.

It signals at least a subconscious awareness of two revolutionary concepts underlying the realization of Real World Awareness:

- The migration of IT from the nebulous virtual world of the Net into chips implanted directly into items of clothing, and even our pets
- The automation of IT and increasing linkage of networks and databases

No matter what the benefits and value are to consumers, the adoption of Real World Awareness, particularly in our homes and ourselves, will prove no doubt to be an unsettling experience. If moving into a landscape driven by Real World Awareness could be compared to sliding into a hot bath, consumers are now at the moment where they wince at just how hot the water seems.

The challenge, then, for businesses hoping to pioneer Real World Awareness technologies is to reassure and educate consumers and to humbly concede that their fears of privacy invasion and future, unintended consequences of Real World Awareness proliferation are not entirely unfounded.

That said, the news is encouraging. According to an October 2004 survey by the marketing intelligence company BIGresearch, 28.2 percent of more than 7,000 people polled know what RFID is and could explain how the technology works, and a majority of those who are aware of RFID think that it is a "good thing." Follow-up discussions with some poll members in a focus group setting revealed that those who were aware of RFID have a balanced view of its benefits versus potential privacy issues.

The flip side, however, is that two-thirds of those polled said that they are either "somewhat concerned" or "very concerned" about the potential for privacy abuse, and those who were previously unaware of the technology tended to gravitate toward the privacy issues rather than the benefits.

The sooner consumers are educated about RFID's benefits and potential value, the better it is for everyone.

THE PRIVACY DEBATE

Consumer advocates are most concerned about the potential of RFID to invade individual privacy. This section explores this issue from a number of angles.

There are a number of unresolved issues in the privacy debates raging around Real World Awareness between advocacy groups, like CASPIAN and the EFF; technology boosters, like the think tank the Progressive Policy Institute (PPI), which recently published an issue-by-issue counterattack against RFID's critics; and standards bodies, like EPCglobal, which are attempting to devise a set of self-regulating principles to ease consumer worries and head off legislation that could retard widespread adoption.

WHAT PRIVACY ADVOCATES FEAR

In November 2003, CASPIAN, the EFF, the ACLU, and a large number of co-signing civil liberties groups published their position paper outlining the potential threats to personal privacy contained within Real World Awareness technologies like RFID and their proposed guidelines for use by governments and corporations alike. The paper ended with a call for a voluntary moratorium on item-level tagging until a complete set of usage guidelines can be drafted for what is essentially an immature technology.

As extreme as that particular position is, many of the individual arguments have merit and must be understood by IT managers hoping to work with, rather than against, their employers' own customers. These groups took particular issue with these topics:

- **In-store tracking of customers.** The paper called for an end to RFID-enabled smart-shelf pilot projects in which consumers who picked certain products off shelves at a participating retail store were photographed by surveillance cameras.
- **The potential for hidden tags and readers.** The tiny size of EPC-standard RFID tags, coupled with the lack of line-of-sight restrictions for readers, creates a ripe environment for surreptitious scanning and tracking. RFID tags sewn into clothing or built into purses, shopping bags, and suitcases, for example, have great potential for privacy violations, as do readers hidden under floor tiles, woven into carpeting and floor mats, hidden in doorways, and built into shelving or counters. In each case, it would be impossible for consumers to know whether or when they are being scanned.
- **The potential for massive data aggregation.** Ubiquitous EPC-compliant RFID tags would become the mechanism for creating a global item-registration database in which nearly every physical object can be tracked and that could, thanks to database sharing, be used as a starting point to spy on and track consumers and political dissidents, for example.
- **Rogue tracking.** The imminent proliferation of RFID readers, most likely as the result of cell phone manufacturers' adding reader functionality to next-generation phones, would lead to a scenario in which privacy invasion by individuals is constant.

MYTHS SURROUNDING RFID

Real World Awareness supporters have tried to downplay these fears, primarily by emphasizing how RFID and similar technologies do not have, and may never have, the capabilities required for the massive privacy invasion envisioned.

This list debunks some of the issues highlighted by RFID critics:

- **The scanning range is extremely limited.** Scenarios that depict criminals or FBI agents parked outside homes with RFID readers fail to take into account that EPCglobal tags are passive and operate at a UHF frequency that has difficulty interacting with metal and water. Tags next to human bodies (which are composed primarily of water) are often unreadable. Surreptitious scanning of someone's person or home would require that the spy in question get up very close and personal with a handheld scanner, and be *very* subtle about it.
- **Readers will not be ubiquitous.** Real World Awareness implementation projects are not built around the idea of implanting readers every few feet so that a pallet or individually tagged object is more or less continually scanned. Rather, readers would be placed only in strategic locations—warehouse doors and smart shelves, for example.
- **Tags can be encrypted or otherwise secured.** Crucial tag data is not, and never will be, an open book for anyone with a reader. Tags are not foolproof—no data ever is—but as with Internet commerce, improved encryption will pave the way for an environment in which almost all data can be considered secure for everyday use. As with the Internet, it will be up to consumers to decide whether they can accept the remaining risk.
- **Corporate databases are not shared by competitors or even noncompeting companies in different industries.** The assumption that a retailer would open its customer database to its competitors, or even a broader audience, flies in the face of business logic. Privacy advocates' fears of massively aggregated databases are not realistic. Many laws around the globe—including the Fair Credit Reporting Act, and the Gramm-Leach-Billey Act in the United States, along with the European Union's general data protection directive and electronic communications directive—regulate the use and reuse of consumer data aggregated from bar codes and credit cards. As for fears of government surveillance, every existing medium, and every new medium to come, has the potential for being an avenue of government surveillance. Strict policies, such as the necessity to obtain a warrant before instituting a wiretap, already regulate this potential usage.

However, consumers cannot rely on the current limitations of Real World Awareness technology to protect them from potential abuse. At the very least, an agreed-upon body of standards and practices must be put in place to notify consumers of the presence of RFID tags and other technologies and to enable them to choose whether they will accept the use of tags in everyday settings.

WHAT DO CONSUMERS DESERVE FROM REAL WORLD AWARENESS?

Everyone—critics, proponents, and manufacturers alike—agrees that guaranteeing information transparency is of paramount importance to consumers and is the appropriate starting point from which to draft a bill of consumer rights.

What transparency means, however, depends entirely on which party in the debate you ask. Still, there are some principles upon which both EPCglobal—which has drafted a set of guidelines that it expects licensees to follow starting January 1, 2005—and privacy advocates such as CASPIAN agree:

- **Consumers must know when a tag is present.** The use of tags in products or packaging should never be hidden from the consumer. In the EPC implementation, an EPC logo affixed to the product will notify the consumer that an RFID tag is present. EPC has committed to always making its usage guidelines publicly available and expects its partners to do the same.

- **Consumers are allowed to discard tags, and should know that they are allowed to do so.** "Consumers will be informed of the choice that they have to discard, disable or remove EPC tags from the products they acquire," according to the company's guidelines. How this will eventually be done is still up in the air: The Gen 2.0 specification of the EPC standard will also include the ability to "kill" tags at the point of purchase, by rendering them blank and unusable. Another possibility is "blocker tags," which interrupt another tag's signal temporarily—in effect, render it invisible. (Blocker tags have been tested but not yet been put into production.) EPCglobal also assumes that tags will be embedded in packaging, not products, and will usually be discarded after purchase.

The pro-technology Progressive Policy Institute has issued its own set of recommendations urging that EPCglobal work with the Federal Trade Commission to create an industry-friendly set of usage standards. They maintain that consumers should be notified that tags are present, but written consent should not be required and neither should the automatic killing of tags at the point of purchase.

In the coming months, an open debate between consumers, politicians, companies, and standard bodies is required to come up with solid guidelines to ensure that consumers are informed about the placing and use of RFID technology.

REAL WORLD AWARENESS AND THE LAW: U.S. PERSPECTIVE

In some countries, opponents of Real World Awareness technologies are mobilizing government support against item-level RFID tagging, and have already introduced legislation. The state legislatures of California and Utah have already rejected bills, introduced in early 2004, that would have required the strict notification of consumers by retailers when RFID tags are in use.

The possibility of patchwork laws governing RFID use (similar to legislation passed in some states in the United States in the 1970s, when bar codes were first introduced) is a cause for some concern with the technology proponents.

Such laws could negate some benefits by preempting a streamlined RFID rollout.

A TRADITION OF PRIVACY IN THE EU

In the European Union, member states grappled with electronic privacy issues as far back as the 1970s and emerged with a comprehensive body of law with strict regulations designed to protect individual rights. More important, however, is that these laws are clearly defined and that Real World Awareness technologies fit into the existing legal framework of most countries, by eliminating the potential for confusion and unexpected legislation down the road.

In Germany, for example, the 1977 Federal Data Protection Law created the precedent for laws like the Information and Communication Services (Multimedia) Act, which firmly established legal protection of information used by computer networks. These and other laws were later revised in order to harmonize with the European Union's Data Protection Directive of 1995, which established union-wide policy.

WHEN WILL THE TIPPING POINT ARRIVE?

The debate outlined in the preceding section has everything to do with item-level tagging, which at the moment is assumed will happen sometime this decade, but it is still largely theoretical. The reality is that item-level tagging has almost nothing to do with industrial assets and pallet- and case-level tagging, which comprise the bulk of most pilot projects and implementations to which even CASPIAN is on the record as having no particular objections.

The argument that RFID and, indeed, all Real World Awareness technologies are still immature may be a case of understatement. The question is, "When will the tipping point for item-level tagging—and, thus, wide adoption—actually arrive?"

THE MAGIC NUMBER: $.05

Considering the potential cost of the billions of tags needed annually for effective item-level tagging, the magic number that would herald widespread implementation would appear to be five cents per tag. That, analysts agree, is the magic number at which the potential ROI falls into line with the costs of implementation. But analysts cannot seem to agree on when, if ever, RFID tags will cost that little. Although some point to promising, but experimental technologies being developed by start-up chipmakers, others insist these advances ultimately won't make an impact. For the most part, prognostications have tended to follow regular cycles of pessimism and optimism.

In February 2004, Forrester Research issued the gloomy report "Exposing the Myth of the Five Cent RFID Tag." Although vendors like Texas Instruments will ship 7 billion units per year by 2008, the report said, the average cost of EPC Class 0 or Class 1 tags (the most common and least expensive types) will fall to just .26 euros by 2012, Forrester concluded. Tag assembly (the attachment of an antenna to the chip) would still comprise 35 percent of the manufacturing costs and would partially negate advances being made in chip fabrication by startups like Alien Technology.

A few months later, the picture had improved somewhat. By August 2004, research firm ARC Advisory Group issued its own report concluding that EPC-compliant tags would hit 16 cents by 2005. Better, but not good enough: The absence of a five-cent tag would continue to inhibit adoption, in the ARC analysis.

And it has. According to poll results announced in October 2004 by another analyst firm, ABI Research, the number-one worry holding back RFID adoption at the 50 companies polled was ROI uncertainty (23 percent). Tag and transponder costs came in third with 13 percent. The second-biggest concern was technology standards, discussed next.

UNCERTAINTY ABOUT STANDARDS

Although EPCglobal has gained significant traction as the potential de facto standard for RFID tags (and perhaps for almost all Real World

Awareness technologies to come), EPCglobal itself has issued six drafts of tags aimed at supply chain customers alone. Each tag, ranging from Class 0 to Class 5, has a different intended purpose and different capabilities, with escalating costs for higher-numbered classes. Only Class 0 and Class 1 are now manufactured in bulk, which leaves potential customers to wonder which, if any, of the other classes might be the best option for their particular business process needs further down the road.

Further complicating matters is the continued existence of other tag makers still struggling to supplant EPCglobal as the industry's de facto standard. Companies that continue to produce proprietary tags or identification schemes for customers needing closed-loop solutions are in essence slowing adoption of a broader, probably EPC-based standard.

POSSIBLE SOLUTIONS AHEAD?

More optimistic RFID customers can also take heart from the rapid pace of innovation at chipmakers like Texas Instruments and upstart companies like Alien Technology and Ionic Fusion Corporation, the last of which is getting into the fabrication game by using a home-grown technology process to fuse chips and antennas, thus lowering the cost of assembly.

In October 2004, Texas Instruments meanwhile announced, a new manufacturing process that it claims will enable its partners to assemble Class 2 chips at the magical five-cent mark by 2006 and enable Texas Instruments to produce billions of such units by 2006, as long as the demand exists.

When will the five-cent tag arrive? That would appear to depend on which analyst firm you choose to believe.

WHAT THE FUTURE HOLDS FOR REAL WORLD AWARENESS

This chapter began with the observation that the short-term impact of new technologies is always overestimated and that the long-term impact is always underestimated. Driven by RFID mandates on the one hand and new, upcoming manufacturing processes for RFID tags on the other, in the case of Real World Awareness, the short term may well give way to the long term much sooner than anyone would have anticipated.

Corporations undoubtedly are being exposed to supply and demand pressure to embrace the technological and business process opportunities afforded by Real World Awareness, opportunities that now appear modest—supply chain enhancements, preventive maintenance, authenticated pill bottles—but opportunities that promise to expand in scope at an unimaginable pace.

In the long term, the virtuous circles formed by the linkage of Real World Awareness to other advancement in IT (especially the process integration abilities promised by Web services and an service-oriented architecture) will spawn new applications that have not even begun to percolate in their inventors' minds.

If there is one thing, though, that this chapter hopefully made clear, it is that only applications designed with the consumer firmly in mind—applications that drive value for consumers while continually reassuring them that this value absolutely outweighs any concerns—will survive. The success of such applications depends on widespread adoption, which will never happen if the consumers are dismayed, put off, or perhaps even frightened, rather than eager for the promised benefits.

NOTES

1. Lytel, D. "RFID: Investing in the Next Multi-Billion-Dollar IT Opportunity," *Precursor*, November 12, 2003.
2. Overby, C.S. "Revealing RFID's Benefits in Consumer Goods," Forrester Research report, October 12, 2004.

PART II

EXPERTS IN REAL WORLD AWARENESS

7

Mastering the Legal Challenges

The Damstadt University of Technology is one of the leading universities in Germany. In 2002, Professor Viola Schmid took on the chair for Public (International) Law and Information Technology Law. Research areas are national and international cyberlaw, information technology law, and data privacy law. Areas of special interest are the challenges caused by pervasive and ubiquitous computing and the law of equity in cyberspace. Viola

works in the project SicAri—a Security Architecture and Its Tools for Ubiquitous Internet Usage, funded by Germany's Federal Ministry of Education and Research.

Viola Schmid practiced business law at Fries Lawyers, Germany, and holds a master's degree in law from Harvard Law School. She joined the Free University of Berlin as a lecturer of public law and taught courses in German, European, and energy law. Her publications (doctorate and post-doctorate degrees) focus on transnational commercial speech doctrine, environmental and energy law, and cyberlaw. Viola was born July 2, 1960, in Augsburg, Germany. She is married and has two children.

Prof. Dr. Viola Schmid[1]
LL.M. (Harvard)

193

Radio Frequency IDentification provides the opportunity to improve products and services, make them safer and cheaper, and even protect people and animals from themselves and others.[2] The use-oriented chapters of this book express the hope that "RFID will revolutionize . . . the way we do business, and deliver unimaginable benefits."[3] Two main legal arguments can be made against making use of these numerous opportunities:

- The right to *data privacy,* which is the right to privacy with respect to the collection, processing, and storage of personal data by automatic means.[4]
- The right to *data security,* or *IT security,*[5] which is the existence of reasonable security safeguards protecting personal data from such risks as loss or unauthorized access, destruction, use, modification, or disclosure.[6]

SOME RFID SCENARIOS

These rights, which are concepts of European and public international law, are now on the verge of becoming new legal issues in the American legal system. Paradoxically, an inverse relationship exists between these rights and the benefits of RFID: "In order to have the most value to both individuals and society, the infrastructure (to read tags) needs to be widespread. . . And yet it is just the widespread infrastructure that raises the most questions."[7] Many of the RFID applications that promise the greatest technological, medical, and economic benefits also create the greatest legal challenges. This chapter examines three of the many possible scenarios.

EPC SCENARIOS

Electronic product codes (EPCs)[8] are a strategy for the real-time enterprise (RTE)[9] and contain, for example, the following type of information: "The cola can was produced at the New York plant on September 9, 2004." This is EPC scenario 1; it doesn't involve personal data and therefore has no relation to existing privacy provisions of the United States Constitution and the United States Code. EPC scenario 1 can then theoretically be combined with two other data records, to create two more scenarios. EPC scenario 2 includes details about the sale: "The cola can was sold for U.S. $1 on October 6." EPC scenario 3 involves the customer: "bought by credit card holder X." The combination of

EPC scenarios 1, 2, and 3 triggers privacy protection under U.S. and European law because "personal data"[10] are involved.[11]

The reasons for employing EPCs are at least twofold: The cola example is part of an asset management strategy, whereas the federal Food and Drug Administration recommends RFID tagging for drugs as a means of combating counterfeit drugs: "In recent years, however, the FDA has seen growing evidence of efforts by increasingly well-organized counterfeiters backed by increasingly sophisticated technologies and criminal operations to profit from drug counterfeiting at the expense of American patients."[12] The impact of counterfeit drugs on the health, liberty, or happiness of humans can be devastating. RFID may offer a welcome remedy to minimize these risks.

RTAMP SCENARIOS

RTAMP, or the real-time authentication and monitoring of persons, occurs when access authorizations need to be checked wirelessly or when RFID tags distributed around the home are used to determine whether elderly people requiring care have, for example, taken their medication, brushed their teeth, or eaten (Activities of Daily Living-Monitoring, a visionary project of Intel that was presented at a workshop of the Federal Trade Commission).[13] Another scenario is the attachment of RFID to students' school bags or nameplates while tag readers are installed at the school gates and at locations the students' parents and teachers think could be dangerous.[14] In 2004 the implementation of RFID strategies at a primary school in Osaka, Japan, is the reaction to a 15-minute rampage of a mentally disabled person who stabbed eight children and seriously wounded 13 others in 2001.[15] Reportedly similar strategies are employed at a Buffalo school in a "gritty neighborhood."[16]

RTAMA SCENARIOS

An example of real-time authentication and monitoring of animals (RTAMA) is the planned legislation in Idaho in response to the challenges of bovine spongiform encephalopathy (BSE, or mad cow disease). The aim of the legislation is to make the import of cattle dependent on RFID identification.[17]

From a technological perspective, EPC applications are real scenarios and RTAMP and RTAMA are partly future areas of use. The newness of RFID in everyday use and the diversity of areas of use prevent a final legal assessment at the moment. Yet this newness and diversity require the first steps toward a legal assessment to be made. And these steps are required by not only American or European law and legal theory, but by laws everywhere: A global technology such as RFID will require global RFID law—at least a global discussion of whether any RFID law is needed. The economy thrives to globalize, the products and services will be marketed globally (via the Internet), and the RFIDs attached to products and included in services would be a chill and a hindrance for these market chances if certain national legal systems would object to RFID applications and ban them. But even in this hypothetical scenario, it is evident that information about RFID and deliberation is needed. Even a hypothetical State S with the highest data privacy standards imaginable that banned RFID would be in the greatest danger that its hypothetical anti-RFID laws would be circumvented or without force because RFID will be so widespread.

If steps on a legal stepladder have to be considered, why not consider as the first step in State S the German law? German law was not only a pioneer of data privacy and data security,[18] but it is now also very important for those for and against RFID. RFID is being piloted and rejected on both sides of the Atlantic, as the paradigmatic scenarios in Rheinberg, Germany, demonstrate. On June 23, 2004, the Washington Post reported: "A store in Rheinberg, Germany, took RFID tags out of its loyalty cards after protests. Many large firms working with RFID now have extensive disclosure statements on their Web sites."[19] The article reported on a workshop, Radio Frequency Identification: Applications and Implications for Consumers,[20] on June 21, 2004, organized by the Federal Trade Commission (FTC). At the workshop, Katherine Albrecht, who founded the consumer rights movement CASPIAN,[21] boasted that ". . . CASPIAN uncovered the scandal and rocked Germany."[22] What had happened? Two scenarios can be distinguished: the METRO scenario and the CASPIAN scenario.

METRO Group Scenario

The real-time enterprise METRO, of Germany, has a "future store,"[23] the purpose of which is to test Real World Awareness (RWA) strategies and familiarize customers with such strategies. The METRO loyalty card contained an RFID tag and was designed to ensure, among other things, youth protection in the multimedia department. This department gives customers the opportunity to try out movies before they buy them. They hold their loyalty cards in front of a reader and can then watch selected sequences from the movies—as long as they are at least 16 years old, as required by the German Youth Protection Act. METRO handed out loyalty cards only to people who were at least 16. The reader installed in the multimedia department needed only the customer number: The presence of a loyalty card automatically meant compliance with the Youth Protection Act. As the subsequent events in the CASPIAN scenario showed, this one-way, read-only transfer of data without reference to personal details (apart from the customer number)—totally unspectacular from the point of view of data privacy law—might not have offered adequate data security.

CASPIAN Scenario

On January 31, 2004, Ms. Albrecht received a METRO loyalty card after she and some other activists had been on a visit there. Without METRO involvement, the activists purported to read the tag's memory during a public presentation of Ms. Albrecht's the following day using a RFID reader from the company Megaset. They added the sentence "Thank you, Katherine" to the memory.[24] The CASPIAN scenario is thus a data security law scenario provoked by activists. It was not METRO that illegally read personal or other data but, rather, the activists who accessed the tag's memory. Whatever the legal significance of the activists' reading or writing strategy,[25] there is no dispute that the METRO tag is unprotected against RFID readers and writers. This could be a breach of German data security law,[26] which requires that the storage of personal data by automatic means must be protected against unauthorized access, destruction, use, modification, or disclosure. Why should a separate law apply if an individual's nonpublic personal information is stored in an RFID tag rather than on a handheld or personal computer? Under German law, data privacy requires data security.

RFID QUESTIONS

Should consumers be notified if companies use RFID? Should RFID be covered under data security law? And, will RFID invade consumer privacy? This section looks at each of these questions in turn.

QUESTION 1: DO CONSUMERS NEED TO BE NOTIFIED OF RFID USE?

Here is the clear answer to this question in areas under U.S. jurisdiction: In July 2004, there was not yet any legislation requiring notification.[27] However, some differences already exist in the legal developments between federal and state law, as described in the two following sections.

Federal Level: The Duty of Information

The activist Ms. Albrecht is demanding a duty of information at the federal level with the proposed legislation CASPIAN—RFID Right to Know Act of 2003.[28] The CASPIAN initiative wants the duty of information to also cover RFID that acts purely as a bar code, as described by EPC scenario 1, earlier in this chapter—that is, no personal data is involved, unlike the EPC combination scenarios 1, 2, and 3.[29] All uses of RFID technology should be clearly labeled and indicate at least the following information: ". . . at a minimum, that the consumer commodity . . . bears an RFID tag and that tag can transmit unique identification information to an independent reader before and after the purchase."[30] The proposed legislation is not far removed from the self-regulation policies of some RTEs in the electronic product code industry. They propagate comparable guidelines, "which are based . . . on industry responsibility, providing accurate information to consumers and ensuring consumer choice."[31]

State Level: RTAMA and Duty to Information

The states differ in that one state uses RFID in its legislation as a control instrument while others are occupied with the right-to-know issue.

An example of a state using RFID as a control instrument is in the RTAMA scenario in Idaho. The data privacy interests of the breeders show that RTAMA can change economic reality and the market: "The tags contain medical history, lineage, and price, which livestock owners are wary about releasing. . .We think it's very important to protect that data, and we will not go to a mandatory system until we find a way to protect that data."[32]

The proposed legislation in Utah, Missouri, and California provides examples of the right-to-know issue. In Utah, the first attempt failed in March 2004 because of resistance from retailers who felt too restricted in their RFID plans. The politician introducing the legislation then announced a further breach.[33] The legislative bodies in California and Missouri are still deliberating.[34]

To summarize, the United States does not (yet) have any legal requirement to label goods that have RFID tags. Even nonlegal guidelines for industry responsibility, however, should advise the industry to inform the public. Consequently, the following sentence will not need to be uttered in the future: "RFID industry is in a crisis, but it's not a crisis of functionality—it's a crisis of confidence." The area in which RFID is being used should determine whether the information is provided in the form of labels, notices in stores, or information on a Web site. This chapter thus distances itself from the CASPIAN initiative, which demands labels (with tags attached under the labels) and recommends that the issue of whether the duty of information to the general public is necessary in all RFID scenarios needs to be discussed. There is no apparent legal requirement, for example, that a car thief be warned about the use of an antitheft device. Should an unauthorized person be informed of efficient RFID security applications[35] or a child or an elderly person requiring care be informed of RFID tags (designed to protect her) if the tags then cease to be effective? The RTAMP scenarios let us foresee interesting legal and philosophical discussions about RFID and self-determination, freedom, privacy, and security.[36] In addition to the proposed detailed and use-oriented legal examination, in some instances it would be a good idea from a technical point of view to indicate RFID strategies by using labels. For example, if the channel for transferring data is short and the product to be read is large, a label indicating "RFID inside" helps locate the tag and thus speeds up the read process.[37] To conclude our discussion of the theoretical German state, State S, the German chief privacy officer demands RFID legislation, but, so far, the German government sees no necessity for initiating the legislative process.[38]

QUESTION 2: WHAT REQUIREMENTS ON THE USE OF RFID ARE NEEDED UNDER DATA SECURITY LAW?

The protection goals that are familiar from IT security law—such as identity, authenticity, integrity, obligation, and confidentiality[39]—also apply to RFID. U.S. IT security law requires examination, particularly in the area of technological standardization,[40] which needs to be further developed for specific areas of RFID use. The METRO and CASPIAN scenarios prove the great significance of data security for the deployment of RFID. The Idaho RTAMA scenario also shows you that access rights must be effectively secured, for example, if the food agency should not be granted access to all information that is important to the life of the cattle. The general rule needs to be that the more personal or economically relevant the data processed in the memory, sensor, and logic device, the more specific the rules must be that determine which read devices can (exclusively) access it and using which authentication. This statement is particularly true for RTAMP scenarios and even more so for complex cases within these scenarios, such as improving the care of seniors or students through monitoring. Monitoring can take place only with exclusive access rights—otherwise, there is a risk that people's privacy will be invaded.

The EPC applications occupy a special position: In contrast to the comparatively complex RTAMP and RTAMA applications, electronic product code applications need to be read by as many readers as necessary (or possible). Here, the rule is that RFID can work only without authentication and identification.[41] This "lack of IT security" is justifiable if only EPC scenario 1 is used. Combinations involving EPC scenarios 2 and 3 involve personal data and are thus subject to higher security requirements.

Practitioners point out that the demand for RFID that is secure under data privacy and data security law (the "smart" RFID tag approach) conflicts with the need for cost effectiveness: "With a budget of five cents, there is very little to spend on additional logic gates."[42] The response is that cost effectiveness has always presented a challenge in the quest to make IT secure and sustainable and to improve its functional quality.

Beyond the unilateral, specific security requirements, which depend on the type of application and the complexity of the tag (for example, storage capacity, type of logic device, direction of data transfer—read only or read/write, for example—or distance of data transfer[43]), future RFID law will face another challenge: the assessment of tools that privacy activists use in bilateral and multilateral scenarios to protect their own data. Such "DataPrivatizers" can range from RFID detectors to active jamming approaches that stop the wireless interfaces from working.[44] Depending on how well the tools function and how much they are used, the main deciding factor will be whether, first, society and, second, the legislators, authorities, and courts find the right balance between RFID manufacturers and users on the one hand and consumers— who either accept or reject RFID—on the other. This process will entail considerable effort, particularly for EPC RFID applications, as the planned legislation in California, Utah, and Missouri illustrates. The social and legal issue will be the extent to which EPC RFID tags can and should be read ubiquitously and pervasively after the business transaction has taken place.

In summary, because there is still no legislation shaping the requirements for RFID use, the detailed legal security criteria are fuzzy and they are difficult to determine—even for informed, motivated early adopters. However, what should become clear is that the total disregard for security arguments, such as in the METRO scenario, predictably leads to a lack of acceptance. Technology needs to be accepted if it is to be successfully marketed and thus have a chance to contribute to the good of the public.

QUESTION 3: IS THERE A DANGER THAT RTES WITH RWA STRATEGIES (IN PARTICULAR, RFID) DIGITALIZE PEOPLE AND THEIR BIOGRAPHIES AND MISAPPROPRIATE DATA?

The answer to this question is clearly Yes. However, there are always risks in life. All technologies come with opportunities and risks, and it is society and the law's task to recognize and analyze these opportunities and risks and, if necessary, develop strategies to align the opportunities and risks optimally with each other. The proposed CASPIAN legislation would add the provisions in the following sidebar to the Privacy subsections of the United States Code.[45]

The background to this demand to create consumer privacy RFID legislation and the responsibility of the Federal Trade Commission for monitoring and issuing data privacy and data security standards is the fear that RFID will further increase the danger that data will be collected and sold in a quantity and quality unimaginable until now. There are many instances of the illegal sale of data without the consent of those affected, as Ms. Albrecht has found in her research: "There are also a number of disquieting cases where Internet companies reneged their privacy policies during hard times by attempting to sell customer purchase data to the highest bidder."[46] The risk of breaking the law is equally high for all technologies—but the consequences are more dangerous with technologies that can digitalize even more data, and thus generate more products (EPC combination scenarios 1, 2, and 3), and sell them at lower transaction costs.

§ 6831 PRIVACY PROTECTION FOR CONSUMERS

(a)

(1) A business shall not combine or link an individual's nonpublic information with RFID tag identification information beyond what is required to manage inventory.

(2) A business shall not, directly or through an affiliate, disclose to a nonaffiliated third party an individual's nonpublic personal information in association with RFID tag identification information to identify an individual.

(b) The Federal Trade Commission shall establish appropriate standards for the businesses described in subsection (a) of this section:

(1) To insure the integrity and confidentiality of an individual's records and information;

(2) To insure that RFID tag records do not identify individuals;

(3) To protect against anticipated threats or hazards to the security of an individual's records and information; and

(4) To protect an individual against substantial harm or inconvenience, which may result from unauthorized access to or use of an individual's records and information

Economists would like to answer the question of whether the incentives to misappropriate data are greater or less with more data (because the costs lessen if more data is available). In cyberspace, the following statement applies: There is no way of making good the misappropriation of data. After databases have been outlined and passed to an unauthorized party, there is no guarantee that their content has been deleted.[47] A wrong cannot be made right again. In addition to evaluating the risks of RFID, its opportunities need to be explored. Therefore, everyone involved in the market (manufacturers, users, consumers, and the public) must be open to discussion and be informed as quickly as possible about the opportunities that RWA offers in order to develop a culture for dealing with its use and its risks. Legislation also needs to be discussed even if it does not concentrate on the same issues as CASPIAN.[48]

The necessity of legal regulations must be decided separately for each scenario. See the section "RTAMP Scenarios," earlier in this chapter, for examples of how labeling can reduce the effectiveness of RFID in protecting people from themselves, and see "EPC Scenarios" to see where labeling is a prerequisite for reading tags quickly if the channel for transferring data is short. Rather than a call for a moratorium, this is a global appeal to lawyers to look into RFID. Maybe RFID needs global legislation—as the fight against cybercrime[49] and spam[50] does. A far less satisfactory outcome would be the future intimated by The Economist: "Scaremongering by some privacy advocates, who fear that details of everything they buy will be held on a database and potentially used for nefarious purposes, has made some firms quite defensive about their RFID ambitions."[51]

In summary: All parties concerned should work on the safety, security, and privacy of RFID and—for lack of a strict legal framework—ask the time-honored question: How do I want my data and that of my children to be handled?

NOTES

1. The author would like to thank Ms. Ruth Schadel and Mr. Andreas John for helping with her research.
2. Floerkemeier, C.; Schneider, R.; Langheinrich, M. "Scanning with a Purpose—Supporting the Fair Information Principles in RFID Protocols" (July 27, 2004), http://www.inf.ethz.ch/~langhein/articles/, p. 5. Shows 15 different kinds of purpose declarations for RFID reader queries: access control, anticounterfeiting, antitheft, asset management, contact, current, development, emergency services, inventory, legal, payment, profiling, repairs, returns, and other.
3. Langfort, S. Wal-Mart's global director of RFID, as cited in Washington Post: Krim, J. "Embedding Their Hopes in RFID," June 23, 2004.
4. European Union law: Directive 95/46/EC of the European Parliament and of the Council of October 24, 1995, on the protection of individuals with regard to the processing of personal data and on the free movement of such data, Article 1.
5. IT security and data security are synonymous in this context. In a future legal perspective, there will be in Germany a fundamental right to data security (Article 2 Section 1 and Article 1 Section 1 of the German Constitution) and a so-called "institutional guaranty" of IT security.
6. Public international law: OECD guidelines on the protection of privacy and transborder flows of personal data (September 23, 1980), Part Two, No. 11, "security safeguards principle"). European Union law: Regulation (EC) No. 460/2004 of the European Parliament and of the Council of March 10, 2004, establishing the European Network and Information Security Agency. Article 4 (c) "network and information security" means the ability of a network or an information system to resist ". . . accidental events or unlawful or malicious actions that compromise the availability, authenticity, integrity, and confidentiality of stored or transmitted data. . . ."
7. Elliot Maxwell, chairman of the International Policy Advisory Council of EPCglobal, as cited in Washington Post: Krim, J. "Embedding Their Hopes in RFID," June 23, 2004.
8. www.epcglobalinc.org/about/faqs.html (July 27, 2004). The Electronic Product Code (EPC) is a unique number that identifies a specific item in the supply chain.

The EPC is stored on a radio frequency identification (RFID) tag, which combines a silicon chip and an antenna. After the EPC is retrieved from the tag, it can be associated with dynamic data, such as where an item originated or the date of its production. Much like a Global Trade Item Number (GTIN) or Vehicle Identification Number (VIN), the EPC is the key that unlocks the power of the information systems that are part of the EPCglobal network.

9. Raskino, M. "Driving Out of the Downturn—The Real Time Enterprise," Conn. L. Trib., Vol. 28, No. 47 (December 2002); Kuhlin, B.; Tielmann, H. The Practical Real Time Enterprise. Berlin: Springer, 2005.

10. European Union law: "Personal data" shall mean any information relating to an identified or identifiable natural person ("data subject"); an identifiable person is one who can be identified, directly or indirectly, in particular by reference to an identification number or to one or more factors specific to his physical, physiological, mental, economic, cultural, or social identity [Directive 95/46/EC of the European Parliament and of the Council of October 24, 1995, on the protection of individuals with regard to the processing of personal data and on the free movement of such data. Article 2 (a)] U.S. law: "Personal information" means individually identifiable information about an individual collected online, including (A) a first and last name (B) a home or other physical address, including street name and name of a city or town (C) an e-mail address (D) a telephone number (E) a Social Security number (F) any other identifier that the Commission determines to be the physical or online contacting of a specific individual or (G) information concerning the child or the parents of that child that the Web site collects online from the child and combines with an identifier described in this paragraph (Children's Online Privacy Protection Act of 1998).

11. Schmid, V. RFID and Privacy, Germany, to be published in 2005.

12. Combating Counterfeit Drugs: A Report of the Food and Drug Administration, Feb. 2004, http://www.fda.gov/oc/initiatives/counterfeit/report02_04.html (July 14, 2004).

13. Fishkin, K. Intel Research, Seattle: "RFID for Healthcare: Some Current and Anticipated Uses," a lecture given at the FTC Workshop, June 21, 2004; http://www.ftc.gov/bcp/workshops/rfid/ (July 27, 2004): "Better Eldercare: Activities of Daily Living–Monitoring: If RFID tags are scattered about a house (either from purchase or manually) and RFID readers can detect when you come near those objects, then we can do a good job of inferring ADLs including medication taking."

14. Japan Today, July 8, 2004, "School to put electronic tags on students to monitor safety" http://www.japantoday.com/e/?content=news&cat=4&id=304748 (July 27, 2004).

15. Pagano, A. "Massacre of Japanese schoolchildren provokes questioning of society," August 3, 2001, World Socialist Web site, http://www.wsws.org (July 27, 2004).

16. Scheeres, J. "Three Rs: Reading, Writing, RFID," Wired News, http://www.wired.com/news/print/0,1294,60898,00.htm (July 27, 2004).

17. There is no federal animal-identification system. Hawks, B. "Review of National Animal Identification Plan," hearings of the Agriculture, Nutrition and Forestry Committee—U.S. Senate, http://agriculture.senate.gov/Hearings/testimony.cfm?id=1070&wit_id=3034 (July 27, 2004). 2004 Idaho Bill No. 816, Idaho 57th Legislature—second regular session.

18. The world's first data privacy law was enacted on September 30, 1970, in the state of Hessen in Germany.

19. Krim, J. "Embedding Their Hopes in RFID," Washington Post, June 23, 2004.

20. www.ftc.gov/bcp/workshops/rfid (July 27, 2004).

21. CASPIAN stands for Consumers Against Supermarket Privacy Invasion and Numbering. Compare the CASPIAN Web site: www.nocards.org (July 27, 2004).

22. www.spychips.com/metro/scandal-payback.html (July 27, 2004).

23. METRO Group: "METRO Group startet die unternehmensweite Einführung von RFID": www.future-store.org/servlet/PB/-s/15k9c5za28j5wasfwsu1dp i2yl2125xy/menu/1002256_pprint_11/1088551037081.htm?part=null (July 27, 2004).

24. Heise online news: www.heise.de/newsticker/meldung/print/44237 (July 27, 2004).

25. Under German law, this may be in breach of telecommunications secrecy (Jürgen Müller, Ist das Auslesen von RFID-Tags zulässig? DuD 2004, 215, 217), but is justified by Ms. Albrecht's right to information under data protection and data security law.

26. For example, the annex to the German Data Protection Act, Section 9, if the METRO tag holds "personal data."

27. The author is working on a paper for German law that would consider a duty of information in the German Data Protection Act.

28. CASPIAN, RFID Right to Know Act of 2003: www.nocards.org/rfid/rfid-bill.shtml (July 27, 2004).

29. "There should . . . be a general presumption that Americans can know when their personal information is collected, and to see, check, and correct any errors," Vermont's U.S. Senator Patrick Leahy, at a conference on "Video Surveillance: Legal and Technological Challenges," Georgetown University Law Center (www.leahy.senate.gov/press/200403/032304.html; July 27, 2004).

30. Amendments to the Fair Packaging and Labeling Program (Title 15 U.S.C. Ch. 39 Sec. 1453 paragraph (9) and Title 15 U.S.C. Ch. 39 Sec. 1453; Title 21 U.S.C. Ch. 9 Subch. II Sec. 321; Title 21 U.S.C. Ch. 9 Subch. IV Sec. 343; Title 21 U.S.C. Ch. 9 Subch. V Part A Sec. 352; Title 21 U.S.C. Ch. 9 Subch. VI Sec. 362; Title 27 U.S.C. Ch. 8 Subch. II Sec. 215; Title 15 U.S.C. Ch. 36 Sec. 1333)

31. (1.) Consumer Notice—Consumers will be given clear notice of the presence of EPC on products or their packaging. This notice will be given through the use of an EPC logo or identifier on the products or packaging.

(2.) Consumer Choice—Consumers will be informed of the choice that they have to discard, disable, or remove EPC tags from the products they acquire. It is anticipated that for most products, the EPC tags would be part of disposable packaging or would be otherwise discardable. EPCglobal, among other supporters of this technology, is committed to finding additional cost-effective and reliable alternatives to further enable consumer choice.

(3.) Consumer Education—Consumers will have the opportunity to easily obtain accurate information about EPC and its applications in addition to information about advances in technology. Companies using EPC tags at the consumer level will cooperate in appropriate ways to familiarize consumers with the EPC logo and to help them understand the technology and its benefits. EPCglobal would also act as a forum for both companies and consumers to learn of and address any uses of EPC technology in a manner inconsistent with these guidelines.

(4.) Record Use, Retention, and Security—As with conventional bar code technology, companies will use, maintain, and protect records generated through EPC in compliance with all applicable laws. Companies will publish, on their Web sites or otherwise, information on their polices regarding the retention, use, and protection of any consumer-specific data generated through their operations, either generally or specifically with respect to EPC use.

32. Wilson, M. "USDA Steps Up Efforts to Track Livestock," CNN.com, May 24, 2004.
33. Swedberg, C. "States Seek RFID Laws: State Legislators in Utah and Missouri Have Sponsored Bills That Would Require Retailers to Alert Customers When Goods Contain RFID Tags, RFID Journal, March 16, 2004, http://www.rfidjournal.com/article/articleprint/833/-1/1/ (July 27, 2004).
34. Missouri: S.B. No 867, RFID Right to Know Act of 2004 last action: March 9, 2004, Hearing Cancelled E-Commerce and the Environment. California: S.B. No 1834 last action: June 22, 2004, set first hearing.
35. Another issue is information for authorized parties, which, according to the opinion in this chapter, is required.
36. For example, because the medication or toothpaste is touched but not used or the children avoid RFID readers at dangerous locations.
37. Association for Automatic Identification and Mobility, Standards, March 2004, http://www.aimglobal.org/technologies/rfid/ (July 27, 2004).
38. German law: "Bundesdatenschutzbeauftragter fordert RFID-Gesetz," Heise online news, May 17, 2004, http://www.heise.de/newsticker/meldung/print/47414; and question of the member of the House of Representatives Gisela Piltz and others (House of Representatives materials 15/3025). Comparable to U.S. law: chief privacy officer, Department of Homeland Security, Washington, DC 20528.
39. U.S. law: The term "information security" means protecting information and information systems from unauthorized access, use, disclosure, disruption, modification, or destruction in order to provide (A) integrity, which means guarding against improper information modification or destruction, and includes ensuring information nonrepudiation and authenticity (B) confidentiality, which means preserving authorized restrictions on access and disclosure, including means for protecting personal privacy and proprietary information (C) availability, which means ensuring timely and reliable access to and use of information. (E-Government Act of 2002; § 3542); German sources: Eckert, C. IT—Sicherheit. Konzepte, Verfahren, Protokolle, 2. Auflage 2003, S. 6ff.
40. "ISO/IEC 15408-1:1999 under 4.1.1: Security is concerned with the protection of assets from threats, where threats are categorized as the potential for abuse of protected assets." According to available research, this does not include any provisions on RFID (July 2004). To protect assets: For example, California S.B. 1389, see California Civil Code § 1798.82, which requires a state agency or a person or business to disclose any breach of the security of data.
41. METRO, positioning paper from February 28, 2004: www.future-store.org/servlet/PB/-s/1w8t0r213ahtv1tbkk5ipvv5iz10gnvqu/menu/1002376_11/index.html (July 27, 2004).
42. Juels, A.; Rivest, R. L.; Szydlo, M. "The Blocker Tag: Selective Blocking of RFID Tags for Consumer Privacy," http://www.rsasecurity.com/rsalabs/node.asp?id=2060 (July 27, 2004).
43. For cloning tags when the data transfer channel is not secure. Kelter, H.; Wittmann, S. "Radio Frequency Identification—RFID," DuD 2004, 331, 333.
44. Juels, A.; Rivest, R. L.; Szydlo, M. "The Blocker Tag: Selective Blocking of RFID Tags for Consumer Privacy," http://www.rsasecurity.com/rsalabs/node.asp?id=2060 (July 27, 2004). Heise online news: "WOS3: Prototyp des DataPrivatizer zur Kontrolle von RFID-Tags," www.heise.de/newsticker/meldung/48190 (July 27, 2004). Christian Floerkemeier/Roland Schneider/Marc Langheinrich, "Scanning with a Purpose—Supporting the Fair Information Principles in RFID Protocols," http://www.inf.ethz.ch/~langhein/articles/ (July 27, 2004).

45. Amendment to Title 15 U.S.C. Ch. 94 Privacy. Both existing subchapters (Sec. 6801 ff) contain provisions for the privacy protection of customer information held by financial institutions.

46. From Katherine Albrecht in 2002, "Supermarket Cards: The Tip of the Retail Surveillance Iceberg," 79, Denv. U.L. Rev. 534, 538 with other proof. See also United States District Court, E.D. New York, Re: JetBlue Privacy Litigation, Master File No. 03-CV-4847), a class action complaint for giving a government contractor five million passenger itineraries in 2002 to test an experimental Department of Defense data mining project.

47. Because it cannot be guaranteed that the data is deleted from all end devices and that the unauthorized party has no way of copying it.

48. The German Federal Data Protection Commissioner and a politician from the liberal party are calling for an RFID law in Germany, www.heise.de/newsticker/meldung/ 47743 (July 27, 2004).

49. Public International Law: Council of Europe: Convention on Cybercrime (October 23, 2001), which the United States has signed.

50. Public International Politics: World Summit on the Information Society, Geneva, July 7-9, 2004; http://www.un-ngls.org/wsis-spam-report.htm (July 29, 2004) reports initiatives of the International Telecommunications Union, OECD-Workshops, and of some United Nations Member States.

51. The Economist, June 26, 2004, "The Future Is Still Smart; Technology, Shopping and Beyond."

8

The Impact of RFID on Supply Chain Efficiency

David Simchi-Levi is a professor of engineering systems at Massachusetts Institute of Technology. His research focuses on developing and implementing robust and efficient techniques for manufacturing and supply chains. His work has been published widely in professional journals on both the practical and theoretical aspects of supply chain management. David is the co-author of The Logic of Logistics, published by Springer in 1997. His book Designing and Managing the Supply Chain was published by McGraw-Hill in August 1999. His third book, Managing the Supply Chain, was published by McGraw-Hill in December 2003. He is the founder and chairman of LogicTools (www.logictools.com), which provides software solutions and professional services for supply chain planning. These solutions have been used widely to reduce costs and improve the service level in large-scale supply chains. Clients include Colgate-Palmolive, ConAgra, Del Monte, Kraft Foods, Ryder, SC Johnson, UPS, the U.S. Postal Service, Walgreens, and Weyerhaeuser.

David Simchi-Levi[1]

MIT, Engineering Systems Division, dslevi@mit.edu

R adio Frequency IDentification technology deploys tags that emit radio signals and devices, called *readers,* that pick up those signals. The tags can be active or passive: They either broadcast information or respond when queried by a reader, respectively. They can be read-only or read/write and one-time or reusable. The tags can be used to read an electronic product code (EPC), a unique number that identifies a specific item in the supply chain, and to record information in order to direct workflow along an assembly line or to monitor and record environmental changes. An essential component of the widespread acceptance of RFID is the EPC network, which allows password-protected access to RFID data anywhere in the supply chain.

The proliferation of RFID and full implementation of the technology will take many years to complete. Current forecasts indicate that the EPC network will not be ready until 2007, and it has not yet even been accepted as the standard. In addition, a few challenges remain. These include establishing common international standards for tags, resolving technical problems with tag-scanning accuracy, and reducing the cost of tags. One example is the reliability of tags, which, according to industry analysts,[2] are functioning at success rates of only 80 percent. Antennae sometimes become separated from their tags, and even when the tags stay intact, tag readers are not always reliable. Other problems occur in trying to read tags through metal or liquids and in preventing interference from nylon conveyor belts. Other issues with RFID that have not been resolved relate to policy issues, such as privacy concerns.

Nevertheless, specific mandates by channel masters, such as Wal-Mart, will accelerate the immediate use of RFID, even if it is only at the "slap and ship" level. This technique involves putting a tag on a case or pallet heading out of the warehouse. The tag is scanned at the supplier's loading dock, and a ship notice is e-mailed to Wal-Mart, which compares it to incoming shipments. Even this limited application has advantages in speeded order payment and charge-back resolution.[3]

All these exciting developments have already created a thriving and fast-growing market for RFID technology and services. Many companies are experimenting with various applications of RFID and preparing to comply with mandates from industry giants such as Wal-Mart and the world's largest procurement agency, the U.S. Department of Defense (DoD). These applications include using RFID to improve manufacturing processes, managing SKUs in distribution centers, and tracking products or containers. According to the Wireless Data Research Group (WDRG), the market for RFID hardware, software, and services will increase at a 23 percent compound annual growth rate (CAGR), from more than \$1 billion in 2003 to \$3 billion in 2007.[4]

Most current applications are *within the four walls,* where the benefit from RFID is evident and implementation is relatively easy. The question, of course, is how does one achieve efficiencies beyond a single facility? That is, how can RFID technology be used to improve supply chain efficiencies? The supply chain pundits and technology experts like to say that RFID will increase supply chain efficiency through better visibility and the acceleration of processes in the supply chain. We, of course, do not dispute these ideas, but the meaning is vague; visibility and accelerated processes allow the supply chain to respond better, but what business processes make this possible? Clearly, these processes need to consider not only supply chain complexity but also economies of scale in addition to variability and uncertainty. Indeed, as observed in a recent article,[5] only through a combination of technology and business processes can the supply chain achieve significant improvements.

Thus, the objective of this chapter is to propose a framework and a process for using RFID to improve supply chain performance. Our analysis focuses on the advantage that RFID provides over point-of-sale (POS) data as well as issues associated with supply chain collaboration, technology cost, and who pays for the technology.

RFID APPLICATIONS

Two important drivers now motivate companies to start experimenting with RFID applications. One is the mandate by some major channel masters and procurement agencies, and the second is the collection of immediate benefits that can be gained from implementing the technology.

An important decision that suppliers and manufacturers need to make when considering RFID is level of implementation: by pallet or case or by individual item. Item-level tagging is required in order to achieve many of the benefits of RFID, such as preventing counterfeiting and theft. Unfortunately, because of tag cost, item-level tagging will first be implemented only for high-value items, such as cars and ink cartridges.

Current information systems manage data about products at an aggregated level, such as number of items or cases. Thus, the tracking of single products or cases by implementing RFID requires new information technology (IT) to support it, even at the case level. Of course, the largest benefit can be achieved from implementing RFID at the product level. For example, with RFID, information can be stored in a database about when a particular package of beef was packed, which cow it came from, which farm the cow was from, and where the cow was slaughtered. This type of data can be provided in real time across the supply chain as palettes roll into the warehouse or items roll off the shelves. Modeling this data is a huge technical challenge on its own, and writing applications to understand and use this level of information, or even palette-level information, is a major issue that companies have to deal with.[6]

Some mandated applications within the next few years include the ones in this list:

- **Wal-Mart Stores, Inc.** The top 100 suppliers must tag pallets and cases shipped to three Dallas-area distribution centers by January 2005.
- **Department of Defense.** Pentagon suppliers must place RFID tags on cases or pallets shipped to the DoD by January 2005.
- **Food and Drug Administration.** The organization recommends that all pharmaceutical producers, wholesalers, and retailers begin developing plans to place RFID tags on pallets, cases, and unit items by 2007.[7]

Some applications of RFID already exist, mostly implemented in one facility or one process. The following list describes some examples:

- Package tracking: **United Kingdom Breweries** uses ruggedized RFID tags from a third-party service to track the location and ownership of beer kegs.[8]
- Product tracking: **Michelin North America, Inc.,** has implanted RFID tags on some tires to track their performance over time.[9]
- Store: **Marks and Spencer** has been testing item-level tagging after an initial trial, starting October 2003. Executives have said that the small paper tags enable the company to accurately track its inventory rather than have to deduce the number of specific clothing items in its stores.[10]
- Manufacturing: **Club Car, Inc.,** a Georgia maker of golf cars and utility vehicles, implemented RFID use in the manufacturing process of a new high-end car, named Precedent. The process begins by permanently

installing an active RFID tag on every assembly carriage of the Precedent. At each stop on the assembly line, the carriage passes a reader that sends the car's identifying data to a proprietary manufacturing execution system. The software determines which custom options should be installed in the vehicle and which machine requirements, such as torque, must be completed. Before the car leaves the post, workers make sure that the tasks for the location have been completed. This process replaces the use of printed instructions and cuts the time spent building a new Precedent from 88 minutes to 45. The expenditure for the system was less than $100,000.[11]

- Warehouse management: **Gillette** launched a major RFID trial in 2003, in its packaging and distribution center in Fort Devens, Massachusetts. The company is tracking all cases and pallets of its female shaving systems within this facility. Gillette now knows where every case of Venus razors is in its pack center, how long a case is there, where it is stored, and when it is shipped. The goal of the pilot is to develop systems and business processes needed to sustain extraordinary levels of efficiency and productivity. When the company rolls out the Wal-Mart tagging requirement, it can also eliminate manual case counting, scanning, and other expensive tasks.[12]

RFID AND POINT-OF-SALE DATA

The data commonly used by retailers and their suppliers to forecast demand is *point-of-sale (POS)* data. POS data, taken from cash registers, measures what is sold. Specifically, this historical data is used by many demand-planning tools to forecast demand. Unfortunately, POS data does not measure real demand because it cannot gauge lost sales caused by out-of-stock conditions.

Indeed, a large number of sales are lost because items are misplaced or not on the shelves, where buyers can find them. Lost sales from out-of-stock goods are conservatively estimated at seven cents on the dollar,[13] but the truth is that no one knows the real value. For example, an article by Raman, Dehoratius, and Ton[14] documents many distribution-center and store-level execution problems that lead to customers not being able to find products in stores. Some drivers of execution problems are related to store and distribution center replenishment processes, such as scanning errors, items not moved from storage to shelf, wrong items picked at the distribution center, and items from the distribution center not being verified in stores. A large variety, cramped storage, and high inventory make it hard to maintain accuracy and replenish shelves.

This situation leads to misplaced SKUs and significant discrepancies between physical inventory levels and information system inventory records.

These circumstances create a huge opportunity for RFID, of course, which will provide much more accurate information about available inventory. For instance, companies that comply with the Wal-Mart mandate will receive information (much more detailed than POS data[15]) that includes these events:

- Received at Wal-Mart distribution center
- Departed distribution center
- Received at store
- Departed store stock room (arrived on shelf)
- Case (or tag) destroyed

This information can provide these immediate benefits:

- Better control over overage, shortage, and damage claims management and the ability to better assign responsibility to the supplier, the carrier, or Wal-Mart
- Better control over product recall
- Data that improves processes through collaboration between Wal-Mart and its suppliers
- *For the first time, quantification of lost sales.* This is the true advantage of using RFID information over POS data. Because retailers know what is sold, what is in inventory, and when shelves are not stocked, they can determine *realized* demand based on *actual sales plus lost sales.* This analysis requires new statistical and forecasting techniques that take advantage of the new information.

BUSINESS BENEFITS

RFID implementation will improve both the accuracy and speed of data collection. This accuracy is achieved by a reduction in scanning errors, better prevention of theft and diversion, and the efficient tracking of expiration dates for spoilage. Speed is gained by less product handling and ease of performing an inventory count in a facility through multi-object scanning, for example. Combined with new processes, these factors will lead to an acceleration of the supply chain that results in new supply chain efficiencies.

Retailers are expected to be the main beneficiaries of RFID implementation. According to a study by AT Kearney,[16] retailers expect benefits in three primary areas:

- **Reduced inventory.** A one-time cash savings of about 5 percent of total system inventory. This savings is achieved by reducing order cycle time and improving visibility, which leads to better forecasts. A reduction in order cycle time yields a reduction in both cycle stock and safety stock, and an improved forecast yields a reduction in safety stock.
- **Store and warehouse labor reduction.** An annual 7.5 percent reduction in store and warehouse labor expenses.
- **Reduction in out-of-stock items.** A yearly recurring sales gain of seven cents per dollar caused by fewer out-of-stock items and less theft.

Overall, a retailer with a wall-to-wall RFID system, which includes readers and actionable real-time information that feeds corporate databases, can save 32 cents on every dollar in sales, after taking into account the cost of implementation.[17]

Companies are struggling with estimating the cost of implementing RFID—a cost that is significantly different for manufacturers and retailers. The estimated direct costs for RFID implementation include:

- **Tagging.** A recurring cost incurred by manufacturers. Current tag costs are 20 to 50 cents per tag, depending on volume, and will go down to five cents, according to some contracts, by the end of 2006.[18] Because tagging is done during production, companies will use tags on all cases, even ones going to customers who do not require them. Therefore, the cost for a company that ships 20 million cases per year will be $4 million this year and, eventually, approximately $1 million per year.
- **Readers.** Mostly a fixed cost that retailers and manufacturers will incur. Preliminary estimates for large retailers include $400,000 for a distribution center and $100,000 per store.[19] The only ongoing (variable) cost is for hardware and software maintenance.
- **Information systems.** Results in long-term benefits from RFID (as we describe next) when information systems learn to handle the type of real-time, item-level information that RFID provides.

Manufacturers can also benefit from RFID, depending on the type of business. Immediate benefits from internal implementation include:

- **Inventory visibility.** Better tracking of inventory throughout a company's facilities
- **Labor efficiency.** Reduced cycle counting, bar code scanning, and manual recording
- **Improved fulfillment.** Reduced shrinkage, improved dock and truck utilization, and improved ability to trace products

In the long term, both manufacturers and retailers will benefit from a significant reduction in the *bullwhip effect,* which suggests that *variability increases as one moves up in the supply chain.*[20] Indeed, it is well known that complete visibility throughout the supply chain, such as RFID provides, reduces supply chain variability. This not only allows a reduction in inventory levels, but also leads to better utilization of resources, such as manufacturing and transportation resources. At the same time, reducing the bullwhip effect also benefits retailers because service levels are improved. Indirectly, manufacturers benefit from a reduction in out-of-stock products by retailers. A 50 percent reduction, for example, provides suppliers with a 5 percent revenue gain.[21]

Evidently, for manufacturing companies selling a low volume of expensive goods, such as drugs and general merchandise, the benefits are quite high.[22] On the other hand, for manufacturers of high-volume/low-cost products, such as food and groceries, the benefits received from RFID are not as clear for two reasons:

- These industries already have efficient supply chains through the implementation of a variety of technologies and processes.
- Uncertainty in these industries is relatively small; hence, demand is highly predictable.

These high-volume/low-cost manufacturers are therefore likely to implement RFID at the case and pallet level until the technology matures and the tag price decreases considerably. As a result, RFID benefits, such as preventing store thefts and providing the ability to read customers' shopping carts, will take a long time to materialize.

However, many benefits can be achieved with case-level implementation. For example, Gillette sees the following business benefits from implementing RFID at this level[23]:

- A lower number of pallet touch points, which results in efficiencies and labor savings
- Elimination of manual scanning of cases and pallets
- Elimination of manual case counting
- A decrease in label printing and application
- Shorter time to check orders before shipping
- Improved order accuracy
- Fewer negotiations with retailers over missing products
- Curtailment of shrinkage at distribution centers and warehouses and in transit

- Improved forecasting
- Lower overall inventory levels
- Increased on-shelf availability of products
- Improved customer service levels

Observe that the first eight bullets result from physical implementation and do not require the development of new business processes. However, fulfillment of the last four bullets requires supply chain coordination and new supply chain processes, as discussed in the following section.

SUPPLY CHAIN EFFICIENCY

Information received from RFID systems throughout the supply chain can provide almost instant real-time visibility of inventory and in-transit product status. It can help improve the performance of inventory, transportation, and replenishment systems that rely on this information.

In a supply chain with zero lead time, no capacity limits, and no economies of scale, RFID technology will lead to an immediate supply reaction to every demand instance. Thus, in an ideal supply chain, production and transportation lots are of unit size and the supply chain is managed based on the status of each facility. Specifically, in this type of environment, when a customer removes a product from the shelf, the distribution center ships the product to the retail store and triggers the production of an additional unit.

This is precisely the definition of *lean manufacturing,* in which each manufacturing facility reacts to demand from its downstream facility, in a *pull-based* strategy, and not from its forecast, or a *push-based* strategy. Thus, in an ideal supply chain, the main benefit from RFID technology is the ability to enjoy the benefits associated with lean manufacturing strategies.

Of course, in a real-world supply chain, responding to a demand event is not that simple. First, demand can be replenished from a distribution center, transferred from a nearby store, or satisfied by an emergency shipment from a manufacturing facility. These alternatives provide opportunities to better manage the supply chain, and then they become a challenge as supply chain complexity increases. More importantly, supply chains possess setup time and costs, long lead times, and significant economies of scale in manufacturing and transportation that make a reaction to individual demand triggers impractical.

Therefore, even though RFID provides real-time data, responding in real time to every event is not always a wise decision. Specifically, RFID technology does not imply that it is possible to implement pure pull strategies. So, how can supply chains take into account economies of scale and lead times as they use RFID data? The answer has to do with a hybrid approach in which planning and execution systems are integrated to provide the right balance between the pull supply chains demanded by RFID technology and the push strategies that are required because of lead times and economies of scale. This approach is described in the following section.

CLOSED-LOOP SUPPLY CHAIN PROCESS

How can decision-makers take advantage of the abundance of data provided by implementing RFID technology throughout their supply chains? Similarly, after creating their supply chain plans, how do supply chain executives guarantee that the plans are followed and the assumptions used to build the plan are still valid? Are lead times and lead time variations or customer demands still within a reasonable tolerance of the assumptions that were used? Have supply chain conditions changed and created an opportunity to reset business policies to lower costs or customer commitment times?

Our approach is to develop a closed-loop supply chain process that creates a continuous interaction between planning and execution systems. This approach integrates the execution management of the supply chain with planning to form an effective closed-loop supply chain management system. It requires new technology for both the operational and planning sides.

We present this new concept by using the Total Quality Management's plan-do-check-act process, a cycle of activities designed to drive continuous improvement. This process, first developed by Walter Shewhart, was popularized by Edwards Deming as the *Shewhart cycle*. Initially implemented in manufacturing and with broad applicability in business, this process includes these phases:

- **Plan.** Establish objectives and processes.
- **Do.** Implement those processes.
- **Check.** Monitor and measure processes against objectives and report the results.
- **Act.** Take action to continually improve process performance.

Specifically, this process, when applied to supply chain management, includes the phases described in this list (see Figure 8.1):

- **Plan phase.** The planning system creates a supply chain plan based on data received from current systems about locations, supply chain costs, demand, production—including bill of materials (BOM) structure—and uncertainties in demand, production, and transportation. The supply chain plan includes replenishment and inventory requirements based on current inventory status and manufacturing capabilities. In some cases, a rebalancing of inventory from one facility to another can take place while inventory in others is replenished by building new products.
- **Do phase.** The supply chain plan is transferred to execution systems, and the results tracked by event management systems that also provide visibility and alerts to users. Although local decisions can be taken by users and execution systems, problems beyond local issues are resolved through the planning system.
- **Check phase.** Information about actual demand, transportation, and inventory is fed into a monitor. The monitor, which has access to the results of the planning phase, compares these results to the data from the visibility system and creates various reports and key performance indicators.
- **Act phase.** Based on business rules defined by the user (for example, the gap between the plan and supply chain status), the system can opt to send new data to the planning system to create new plans.

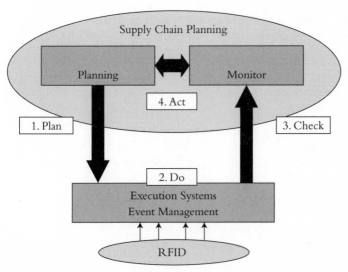

Figure 8.1. The Closed-Loop Supply Chain Process

CONCLUSION

The revolutionary technology RFID will significantly affect the way supply chains are managed and lead to greater efficiency. RFID tags will not merely replace bar codes, but will allow real-time tracking of products, or at least cases and pallets. In particular, RFID will largely reduce lost sales, which is very costly for companies. However, to achieve these efficiencies, new information systems must take advantage of the real-time and detailed product-location information provided by the RFID technology. This chapter has outlined some ideas about how to do this.

NOTES

1. The author would like to thank Edith Simchi-Levi, of LogicTools, for helping him with this project.
2. Bradbury, D. "RFID: It's No Supply Chain Saviour—Not Yet Anyway." *Silicon.com*, September 9, 2004.
3. Rothfeder, J. "What's wrong with RFID?" *CIO Insight*, August 2004, pp. 45–53.
4. Anonymous. "Global RFID Market to Be $3 Billion by 2007." *Supply&Demand Chain Executive*, Oct. 1, 2003.
5. Heinrich, C.; Simchi-Levi, D. "Is there a Link between IT investment and Financial Performance?" Working paper.
6. Bradbury, D., ibid.
7. Rothfeder, J., ibid.
8. Fontanella, J. "Finding the ROI in RFID." *SCMR*, Jan./Feb. 2004, pp. 13-14.
9. Rothfeder, J., ibid.
10. Bradbury, D., ibid.
11. Rothfeder, J., ibid.
12. Roberti, M. "Gillette Sharpens its Edge." *RFID Journal*, April 2004.
13. AT Kearney. "Meeting the Retail RFID Mandate." *AT Kearney*, November 2003.
14. Ananth, R.; Dehoratius, N.; Ton, Z. "Execution: The Missing Link in Retail Operations." *California Management Review*, Vol. 43. No.3, Spring 2001, pp.136–152.
15. Aimi, G. "Finding the Value in Wal-Mart RFID Mandates: It's in the Information." *AMR Research*, June 22, 2004.
16. AT Kearney, ibid.
17. AT Kearney, ibid.
18. Barlas, S. "RFID Bandwagon Rolls On." *Logistics Today*, August 2004.
19. AT Kearney, ibid.
20. Simchi-Levi, D.; Kaminsky, P.; Simchi-Levi, E. "Managing the Supply Chain." McGraw-Hill, 2003.
21. Rothfeder, J., ibid.
22. AT Kearney, ibid.
23. Roberti, M., ibid.

9

Exploding Edges and Potential for Disruption

Professor Charles Fine teaches operations strategy and supply chain management at MIT's Sloan School of Management and directs the roadmapping activities in MIT's Communications Futures Program (`http://cfp.mit.edu/people.html`). *Professor Fine holds an A.B. from Duke University and an M.S. and a Ph.D. from Stanford University. He is the author of many academic articles as well as* Clockspeed: Winning Industry Control in the Age of Temporary Advantage, Perseus Books, 1998.

Natalie Klym is an independent research analyst who has been investigating Internet technology innovation during the last 20 years for a variety of organizations. She is currently a Research Associate at MIT's Computer Science and Artificial Intelligence Lab, conducting research for the Communications Futures Program.

Dr. Dirk Trossen has been with Nokia Research since 2000 and is currently working as a Principal Scientist in the area of seamless and context-aware Internet applications. He received his M.Sc. and Ph.D. from University of Technology in Aachen, Germany.

Charles H. Fine
MIT Sloan School of Management
Natalie Klym
Computer Science and Artificial Intelligence Laboratory, MIT
Dirk Trossen
Nokia Research

I n some cases by design, but mostly by accident, numerous advances in technology have been highly disruptive. The personal computer and the Internet are instructive examples that are empowering new and smaller players to innovate and challenge the existing entrenched economic and technological interests.

The standardization and modularization of the personal computer architecture by IBM were not done to allow many new hardware entrants or to create a desktop software industry. IBM was solving a business problem at the time—and unintended disruptive consequences resulted. The Internet grew from the military and academic research communities. Its openness and decentralized governance structure reflected the values of early users who were interested in sharing information quickly and with little effort. However, a lack of concern about managing and controlling connectivity and the behavior of nodes on the edge of the network led to massive disruption and the creation of many companies that have changed the face of the economy.

The expansion of RFID and other technologies that this book has collected under the umbrella of Real World Awareness seems to be following a much different path from the technologies just discussed. RFID offers a potential explosion of nodes on the edge of the network. Personal computers circle the globe and provide incredible computing power and storage. The hundreds of millions of cell phone chips produced each year provide mobile connectivity along with computing power, albeit less than that of a PC. RFID devices will number in the billions each year, but have much less computing power and memory.

The question of whether RFID will join the personal computer and the Internet in the "Disruption Hall of Fame" is still open. The recent rise of RFID seems carefully choreographed for improving supply chain management, by the big players for the big players. Standards for RFID are amazingly mature for a technology at such an early stage. Large players like Wal-Mart, the United States Department of Defense (DoD), and many others seem to treat RFID as a *sustaining* technology, one that sustains the market power of established players. However, if the big players invest in a new information architecture that enables small players to access new capabilities created by this investment, the stage is set for a disruptive innovator to find a way to leverage the technology in ways not yet imagined by the established players. Just as IBM helped create a network information infrastructure that enabled Dell to disrupt

IBM's industry dominance, we wonder whether Wal-Mart might be speeding its own demise by pushing the adoption of another ubiquitous, but decentralized, information network.

THE GROWTH OF RFID

In our view, RFID technology is poised to enlarge dramatically the global communications network and potentially trigger another wave of value creation through growth and innovation in connectivity. The global communications network has often been characterized as having core servers, switches, and service providers, which provide connectivity for the "mass transit" of information across vast distances. This network core is complemented by devices, applications, and users at the edge of the network, where much value has been created through increased connectivity and communications options. The ongoing expansion of the network's size and capability has been driven significantly by technological innovation in electronics and, especially, semiconductor devices. From central switches and routers that sit at the core of the network to PCs and cell phones located at the network's edge, semiconductor chips are ubiquitous in the communications network. The introduction of RFID chips en masse will expand the global communications network with billions of new edge nodes, with the potential to enable a multitude of new applications and services.

Radio Frequency IDentification (RFID) generally enables the conveyance of information over a relatively short distance. The major driver for developing RFID technology has been the *tagging* of physical objects, striving to enable the "Internet of Things" in which communication and data are associated with and/or triggered by moving physical objects around. With this tagging, particular (physical) object-related semantics are associated with the information that is wirelessly conveyed from the RFID tag to the RFID reader. A typical example of such information is an identifier to be used to uniquely identify the tagged object. The trigger for such information conveyance is typically the close physical proximity of tag and reader. RFID technology then allows for initiating (physical) object-related communication.

Although RFID technology has increasingly become of public interest during recent years, the underlying technology has been developed and deployed since the late 1940s. Examples of the earliest applications include the Friend or Foe (IFF) long-range transponder for identifying military planes during World War II, antitheft electronic article surveillance (EAS) systems in the 1960s, and toll collection in the 1980s. Because the size and cost of microchip technology has shrunk dramatically in the past few years, RFID systems are proliferating and connecting to other data networks and communications technologies.

The first wave of new applications is directed primarily at the supply chain, through the tagging of objects with identifiers to refer to detailed information (related to supply chain management) about the objects. This process bears the promise for huge efficiency gains in management tasks, and many objects outside the supply chain are increasingly becoming targets for similar identification and tracking applications. The driving force of these upcoming supply chain management applications for RFID is expected to drive down the size and cost of tags, which serves the needs of applications that require the mass production of chips (for example, supply chain applications) and those that require microscopic tags.

However, a new wave of RFID applications is expected to drive the notion of the Internet of Things even more; namely, applications that enable communication between devices by exchanging more complex semantics about the tagged physical object than a pure identifier (as in the supply chain scenarios). In other words, rather than tag an object with a code that references a virtual proxy of an object, the chips are integral to a physical object and convey multiple aspects of the object's "state," to trigger semantic-rich communication with another device. For that, more intelligent (and more expensive) RFID chips are required, albeit to enable richer and possibly higher revenue-per-transaction services with RFID technology.

The core-edge working group of the Communication Futures Program at MIT researches the aspect of control in the communication value chain. With respect to RFID, this aspect has been studied by examining the first wave of RFID technology and applications, by focusing on the identification schemes that logically and technically link a tagged object's ID with data describing its properties. The working group's hypothesis is that these identification schemes are a major control point in RFID systems because their design determines who owns and controls the ID data as a crucial piece in the communication value chain for future RFID applications and services.

The question is "How will innovation with RFID proceed, and who, if anyone, will control it?" Right now, it appears that RFID is dominated by large players enforcing their visions and using their market power to compel investments in RFID by smaller or weaker partners, for the primary benefit of the larger players. But these mandated supply chain–oriented investments are just the tip of the iceberg. What would happen if mass customization were possible with RFID tags? What if tags could be easily created and commissioned in every individual home or office by using a device like a printer? What if every cell phone served as an RFID reader? How then would RFID be used, and who would benefit? Would such a development be a gateway to making RFID a disruptive force? Will the current standards-making bodies encourage such developments or try to thwart them? We return to these questions after we clear the decks of the details of how RFID technology works.

TECHNOLOGY OVERVIEW

This section provides a brief introduction to the key pieces of technology needed for RFID deployment: tags and readers.

RFID TAGS

Tags contain a microchip and a transponder. The microchip stores ID data about the object, and the transponder transmits the data to readers. Tags are initially programmed (data is written to the tags) at the point of manufacture (factory programming) but can also be programmed by an OEM or end user (field programming).

Data includes a unique identifier (ID) code based on a standard format and sometimes additional information, depending on the amount of memory on a tag and the application. Sensors may sometimes be integrated with an RFID tag; for example, the tag on an automobile tire may integrate data from temperature and pressure sensors for remote monitoring.

Tags are either passive or active. Passive tags are smaller, about the size of a grain of rice. They are active only when they come within range of a reader's signal. The reader's antenna sends power to the transponder and activates the data stream. Passive tags are much smaller in size and memory and are cheaper to manufacture. Active tags—radio frequency (RF) beacons—are a bit bigger, about the size of a small coin. Each one contains its own power source, thereby constantly transmitting its signal as far as several hundred feet, compared to a passive tag's read range of only a few inches. Active tags can also be rewritten or reprogrammed by readers, whereas passive tags are read-only. Tags transmit over several frequencies, from short-range low frequency to long-range UHF and microwave frequencies.[1] Most passive tags transmit at 13.56 MHz, and most active tags transmit at 433 MHz.

RFID READERS

Readers are the larger, more complex, and more expensive pieces of the RFID hardware (priced at $1,000 or more per reader). Readers are composed of an antenna and a transceiver. The reader captures the information transmitted by a tag, decodes the information, and then delivers it to a host computer for ID resolution and further processing.[2] Readers can be fixed or mobile devices, and some have their own display terminals. Readers may provide *write* capabilities, which means that they can add or change (or *reprogram*) the data on a tag.

Readers have traditionally connected to internal information systems. However, as new network elements, readers are becoming "self-contained computing devices" with a TCP/IP interface that connects them to the Internet.[3]

RFID APPLICATIONS

Today's RFID technology is being driven by supply chain management (inventory tracking) applications. Wal-Mart and the U.S. Department of Defense have set for January 2005 tagging mandates for their top suppliers. But, apart from these rather simple identification-only scenarios, one can envision many more object-related applications.

For this discussion, let's decompose RFID applications into three components: the RFID tag, the RFID reader, and the service infrastructure. We have discussed the first two already in this chapter. The third part refers to the back-end infrastructure needed to perform the desired task for the particular use case. Generally, the RFID reader obtains the RFID tag information via *near field communication,* a wireless communication standard for communications between objects at close range. The service infrastructure then performs an appropriate service based on the information obtained from the tag in order to fulfill the desired purpose. In the following section, we describe several examples in which such services are performed based on the conveyed RFID information.

INFORMATION RETRIEVAL BASED ON SIMPLE IDENTIFIERS

In this application category, information is retrieved that relates to the tagged (physical) object.

The simplest example of such information retrieval is the identification of object owners—for example, to identify the owners of lost (and tagged) pets. Others are the retrieval of product information in a store kiosk, retrieval of information about museum exhibits or public landmarks, and the retrieval of schedules at bus stops.

For these applications, the RFID tag simply conveys a somehow unique identifier that relates to the physical object. The RFID reader delivers the identifier to its service infrastructure, which retrieves the appropriate information. For this, functionality is required about where the information that relates to the physical object resides. The identifier is then used to point to the appropriate piece of information that relates to the particular physical object (for example, the name and contact information of the pet's owner).

INFORMATION RETRIEVAL BASED ON MORE COMPLEX SEMANTICS

Information can be retrieved based on object-related semantics that, apart from some form of identification, is annotated with more complex object-related information, such as state information. Examples of such state information include the reason of failure of a tagged device or generally any kind of sensed information that relates to the tagged object, such as temperature.

An example of information retrieval is device diagnosis for improved customer service.[4]

OBJECT TRACKING

Although the two preceding categories assume that a single RFID tag is usually swiped alongside a single RFID reader for service invocation, one could use a fixed reader network, physically dispersed in a well-known area, for tracking an RFID tagged object. This type of object would travel alongside the RFID readers, and the well-known location of each reader would be used to determine the object's location as well.

Some examples in this category include tracking inventory in a warehouse, tracking inmates in a prison, or tracking children at a theme park.

PAYMENT

In this category, a payment service is invoked based on the conveyed RFID information. Hence, the RFID transaction assumes the fulfillment of a service contract, which leads to a predefined payment transaction.

Examples for these types of transactions are automated payments at highway toll booths and train ticketing.

ACCESS AUTHENTICATION

In this category, the RFID transaction assumes the authentication for (physical or logical) access to information or buildings.

Examples are tagging VIPs for access to restricted areas or tagging club members for access to facilities.[5]

PROXIMITY COMMUNICATION INITIATION

Because a successful RFID transaction indicates the near proximity of two devices, a proximity communication could be initiated between the devices for more sophisticated communication. Additional semantics, conveyed in the RFID transaction, can be used to properly authenticate the communication initiation.

An example of this type of initiation is the establishment of a secure Bluetooth connection between two mobile phones.[6]

ROADBLOCKS

This section outlines a few possible roadblocks to the deployment of RFID-based services.

PRIVACY AND ABUSE OF INFORMATION

Front-end point-of-sale applications are being hindered by privacy concerns, several experiments have been aborted (for example, Benetton and Gillette), and other pilots have met with lukewarm enthusiasm (for example, Prada's multimedia closet). Privacy issues are confining RFID innovation to back-end supply chain initiatives. Consumers are resisting the notion that tags remain active—or even present—outside the boundaries of the store (in the home or in other public spaces). Some retailers see the potential for adding value to products after sale, for example, tire tracking for safety or interactive consumer applications, like tags with care instructions for the washing machine. Nonetheless, consumers remain fearful of invasive "personalized" marketing tactics, unfair price discrimination (detecting consumers with more expensive tastes and pricing their goods accordingly), and remotely monitoring products in the home or cash in wallets by thieves armed with readers. Technologies to "kill" tags when they leave the store are being sought, but some privacy advocates argue that retailers may claim that these items violate the Digital Millennium Copyright Act. On a more general level, privacy issues are increasingly viewed in the larger context of a move toward a "surveillance society."[7]

Privacy issues are also relevant for firms themselves. Ross Anderson has suggested an anticompetitive scenario among geographic markets in the European Union (EU) whereby retailers would discriminate against suppliers on a geographic basis (by detecting the country of origin of products) and thereby violate the EU single-market ideal.[8] Vis-à-vis consumers, RFID *reputation systems,* which enable the cross-indexing of tag codes with consumer-generated information about the firm and the product, may also elicit concerns regarding privacy and abuse of information.

SUPPLY CHAIN SECURITY

Within the supply chain, security concerns include tags that can be damaged, removed, vandalized, falsified, or hacked remotely by consumers or supply chain competitors armed with the proper equipment. The RFDump software, for example, allows tags to be rewritten by anyone equipped with an RFID reader, a laptop or PDA, and a power supply.[9] Tags could be made tamperproof, but because of the way data is written on tags, writing data into the "safe" portion of a tag is more costly.[10]

STANDARDS

Standards issues exist on several levels.[11] Regarding RFID hardware and software, communication protocols are not standardized, and various countries' frequencies are regulated differently across the globe. Hardware standards are necessary for a both a global supply chain and the mass production of tags. Some vendors are working toward producing readers that support multiple protocols and frequencies. [12, 13]

A more complex challenge concerns a standardized naming or identification scheme, which also has multiple levels. Standards pertaining to tag IDs—naming—and a scheme for resolving tag IDs with associated information are discussed later in this chapter.

COST

Tags are still expensive, and implementing an RFID system can cost millions of dollars and take years to deploy.[14] Cost-sharing is a concern, particularly for suppliers who have been receiving mandates to implement RFID from powerful business partners. With the technology still immature and global standards undefined, many companies don't know what kinds of systems to deploy.

COMPLEXITY

System complexity adds to cost issues. Within the supply chain application area in particular, the problem of item-level tagging versus pallet- or case-level tagging is important. Item-level tagging will generate massive amounts of information that must be filtered and managed. Furthermore, applications based on item-level tagging are generally different from applications based on pallets and cases, and they also require different types of tags. And, integrating item-level data will pose additional challenges because existing enterprise systems are not designed for this level of data granularity.

MANAGING THE IDENTIFIERS: AN INTRODUCTION TO RFID ID SYSTEMS

An RFID tag serves as a simple reference to a more complex description of a tagged object's properties. In principle, this model is similar to a library card catalog, where intelligence and complexity are moved from the tag into information systems that store and manage the data associated with the tag ID. (As mentioned earlier, in applications that require the mass production of chips, the size and cost of tags must be minimized.)

Various methods exist for resolving IDs to their associated data. In a very simple application, a tag ID code directly corresponds to a database record pertaining to the object. This type of application tends to use a proprietary identification scheme, in which codes are registered *internally*, with the application provider. Recent examples include pet-tracking applications, like Homeagain and AVID; Snagg's musical instrument authentication application; and NTT's child-tracking service in Japan. We describe these applications in a little more detail.

The Homeagain database is managed by the American Kennel Club, as part of its national Companion Animal Recovery Program. Each ID kit contains an RFID tag programmed with a 10-digit code and a registration form for recording and registering the pet's name and ID number and its owner's name, address, and telephone number. The animal is tagged, the owner registers the pet for a fee, and the information is entered into the Homeagain database by Kennel Club staff. (Although the American Kennel Club controls the database, it is operated by the chip's manufacturer.) Animal shelters, hospitals, and veterinarians (the users) are equipped with handheld wands that read the numbers from lost pets. The vets then enter the numbers manually on the Homeagain Web site to generate contact information.[15] American Veterinary Identification Devices (AVID) is the other main manufacturer of pet microchips. AVID's application uses its own proprietary coding system, and the chip manufacturer (as opposed to a pet organization) controls *and* operates the database.[16]

Snagg basically does the same thing for the authentication of musical instruments. When a customer buys a tagged instrument, or has one retrofitted with a tag, she registers it with Snagg for a fee. Participating instrument manufacturers (like Fender, Gibson, and Carvin) install the chips at the factory (often cross-referencing warehouse serial numbers with Snagg codes). The Snagg database is available to authorized users, including law-enforcement officials, dealers, and repair shops. As in the Homeagain application we just described, ID numbers are scanned with portable readers and the information is entered into an online database.

Child-tracking systems have become increasingly popular in localized areas, like schools and theme parks, and in some countries where child kidnapping is rampant, children are tagged in the same way as pets are. Tagging systems range from wrist bracelets tracked by fixed reader networks to check-in kiosks at designated entry points to subdermal implants. In Japan, NTT Marketing Act Corporation, a subsidiary of NTT West Corporation, offers a child-monitoring system that enables parents to monitor their children from their PCs at home or at the office. The system includes a combination of videocameras and RFID readers that track the movement—or the presence or absence—of tagged name cards worn by children on playgrounds and in classrooms. Data is fed directly over the NTT connection into the application's databases.

Like traditional RFID systems, these applications are all *closed loop*, or self-contained. Each database of objects—pets, instruments, or children—comprises a unique naming context, or namespace, where tag IDs are owned and controlled by the RFID service provider, whether it's the application developer (Snagg, for example), the chip manufacturer (AVID, for example), a national organization like the American Kennel Club (Homeagain, for example), or a telco (NTT, for example).

Other ID schemes are being developed in which the code on an RFID tag tends to function as a *pointer* to sources of information associated with each tagged object—or, in some cases, with each instance of an object. In other words, rather than relate directly to a database record (like the pet's name, owner's name, and contact information in the Homeagain application), the ID refers to the locations of the associated information on the Internet.[17] This type of identification scheme tends to support more complex systems that require interoperability.

In some cases, an object's identity may simply have little to no useful significance outside its own context; few opportunities exist to build services based on open, interoperable coding and resolution schemes. But, increasingly, as RFID applications emerge, tagged objects will have meaning across multiple contexts, especially in the case of supply chain applications, where objects cross multiple organizational boundaries and a single tag ID will be associated with multiple sources of information from various organizations.

Such systems require an ID resolution scheme that resolves a unique tag ID to one or more IP addresses. Because the ID code no longer functions solely within a closed, proprietary system, ID resolution becomes its own, distinct service. IDs are registered *externally;* control shifts outside any single application context, with a neutral third party.

Some RFID applications use URLs as tag IDs, where the DNS effectively functions as the ID resolution mechanism and ICANN as the external registry. URLs are not always the most practical choice for RFID codes because many objects are already subject to widespread, standardized identification schemes, for example, bar codes, passport numbers, and ISBNs. (In combination with XPath expressions, URLs can be advantageous in applications that require tags to convey more complex semantics, but a discussion of these elements is outside the scope of this book.)

Other applications use more arbitrary codes—similar to URLs except that the naming scheme and the registry are not part of the DNS, and are therefore managed by another third party. At this time, the only ID scheme of this sort is EPCglobal's Electronic Product Code (EPC) Network, which uses the Object Name Service (ONS) registry. EPCglobal is a nonprofit organization set up by the Uniform Code Council (UCC) and EAN International, the two organizations that maintain bar code standards.[18] The EPC network, based on technology originally developed by the MIT Auto-ID Center,[19] focuses exclusively on supply chain applications. In addition to tag and reader hardware and software standards,[20] the EPC network relies on standardized codes—EPCs—that are converted to IP addresses by the ONS.

In simple terms, the EPC identifies the manufacturer, the product category (object class), and the individual item (serial number). EPCs are assigned to manufacturers by an EPC namespace authority.[21] Codes are based on the Global Trade Identification Number (GTIN) and complement the existing bar code system.[22] EPCglobal is considering creating EPCs based on other established numbering schemes, like the DoD's unique identification number (UID).[23]

The Object Name Service (ONS) serves as a directory for EPCs. It runs on the same infrastructure as the DNS[24] and resolves EPC-to-IP addresses where information about the object is stored. In addition to a root directory, the EPC network is based on local ONS services, or *caches.* The EPC network uses Physical Markup Language (PML) as a standard for describing objects and all associated information that relates to a particular RFID application. The EPC information service (EPC IS) converts EPC data into the Physical Markup Language (PML) format. Savant middleware runs on edge servers, to gather and filter the data from readers and to interface with the EPC IS and enterprise information systems.[25]

The EPC has gained the support of firms leading the RFID trend in the supply chain—Wal-Mart and the DoD in particular. Nonetheless, adopting the EPC naming system is only a preliminary step and does not necessarily imply or require the simultaneous adoption of the proposed ID resolution system (the ONS). Wal-Mart, for example, may initially use EPCs with its existing EDI-based network and internal information systems (ERP and WMS, for example) to process RFID data.[26] In any case, the EPC network may not be available until late 2005.[27]

More significantly, the EPC Network is only one of several possible naming schemes for supply chain applications and may be challenged as implementations unfold. Forrester reports that both Wal-Mart and the DoD remain skeptical of the EPC identification scheme.[28] In April 2004, the DoD announced that it wants to continue using its proprietary Unique ID (UID) system rather than, or in conjunction with, EPCs, and is working on making it possible to incorporate UIDs on EPC tags.[29] UIDs could be resolved to IP addresses and be less expensive than subscribing to the EPC Network.[30] According to Forrester Research, readers are being designed to handle both EPC and UID tags.[31] One Auto-ID consulting company has also suggested that the DoD could eventually choose to use the IP address of the stored information as the tag ID, based on the new IPv6 protocol.[32] IP addresses, however, are not the ideal identifier because they are not stable, whereas using a code persistently identifies a given object. Furthermore, in complex RFID applications, different instances or states of an object would require multiple IP addresses.

It remains to be seen which solution will be adopted and how it will be authorized. External registries in general are as complicated on a governance level as they are on a technological level, and perhaps more so. Third-party registries don't necessarily guarantee neutrality. ICANN's oversight of the DNS is a case in point. A 2003 research report on online registries notes that "registries can't simply be inserted into the infrastructure and work . . . Registry systems need corresponding infrastructure standards, federated identity management, business protocols and policies, and sometimes even political buy-in or active regulation . . . [M]ost governance questions about active, external registries are unresolved."

Furthermore, as applications develop outside the supply chain, the situation may become further complicated as multiple naming contexts arise for all the "other" things besides supply chain objects. Tag IDs for students, citizens, natural resources, and public landmarks, for example, may be standardized across several categories, each with its own ID type and resolution methods appropriate for its application needs.

Imagine how different the Internet would be if the standards for assigning domain names and Internet addresses were fragmented. Unless some similar scheme is implemented for RFID, it appears that the ID numbers will remain under proprietary control and reduce the potential for mass collaborative applications.

CONTROLLING THE RFID WORLD: A RECOMMENDATION FOR GOVERNANCE

From the perspective of the larger players involved in RFID, supply chain efficiency will improve, as will associated profitability. Instead of being a disruptive force, RFID is being deployed as a sustaining technology that is amplifying the market power of the larger players. The notion of RFID for everybody is unlikely to spring forth from these efforts. EPCglobal is focused on creating an infrastructure for the larger players, and most infrastructure development is focused on the supply chain as well.

But, the potential to tag "everything" and create billions of "objects that talk" is a tempting vision to those who view greater connectivity and ubiquitous network elements as the handmaidens of innovation. At some point, we speculate, a group or a small competitor will figure out a unique "killer" application for objects that talk. Perhaps an RFID tag placed outside a restaurant to contain reviews of its latest offerings will morph into a broader reputation system that can be consulted before deciding to enter all sorts of businesses. Or, will RFID pave the way for services that create the ability to check the total price of a basket of goods across several stores and provide consumers with directions to the best place to shop?

For such applications to become widespread, however, some sort of self-description mechanism or standard will have to be in place for people to make sense of the information. What will that be like?

A centralized standard mechanism like ICANN is unlikely to be developed for RFID. EPCglobal is as close to that as possible, and it will likely stay focused on the needs of larger organizations.

Perhaps grassroots RFID will more likely emerge from usage in the same way that the RSS standard on the Internet was adopted in a collaborative manner. Perhaps the implementation of a broadly agreed-on convention like RSS will take the form of another standard, like web services or UDDI. In this way, an open source collaborative project could develop that would provide a counterbalance to more centralized efforts, such as EPCglobal. Such an open source process is well adapted to the creation of technology to meet shared needs, but also is good at accommodating differences by spinning off related projects. We await such developments with great anticipation.

NOTES

1. For more information about frequencies, see http://www.smartid.com.sg/RFID.htm and http://www.identecsolutions.com/intro_to_rfid.asp.
2. http://www.aimglobal.org/technologies/rfid/what_is_rfid.asp
3. http://www.nwfusion.com/news/2004/070504thingmagic.html?fsrc=rss-wireless
4. The Near Field Communication forum (http://www.nfc-forum.org) envisions a range of such scenarios to use more complex device semantics for improved customer service. For instance, a user might swipe a mobile phone along a malfunctioning CD player to retrieve the identifier of the CD player (for reference purposes), some error code (describing the reason for malfunction), and information on a potential Web page of the customer service. The phone's internal Web browser can use this information to contact the customer service of the manufacturer.
5. http://www.global-elite.org/modules.php?op=modload&name=News&file=article&sid=329&mode=thread&order=0&thold=0; http://www.dailyherald.com/search/main_story.asp?intid=3818786
6. http://www.nfc-forum.org
7. For the "Statement of Barry Steinhardt, Director, Technology and Liberty Project, American Civil Liberties Union, on RFID tags," see http://www.16beavergroup.org/mtarchive/archives/001122.php.
8. http://www.fipr.org/copyright/draft-ipr-enforce.html
9. http://www.eweek.com/article2/0,1759,1628713,00.asp
10. http://www.eweek.com/article2/0,1759,1628713,00.asp
11. See "Roadmap of RFID Tag Standards," at http://www.autoid.org/presentations/rfid_issues.htm.
12. http://www.forrester.com/ER/Research/Brief/0,1317,33298,00.html
13. http://quintessenz.org/rfid-docs/www.autoidcenter.org/about thetech.asp
14. http://www.eweek.com/article2/0,1759,1559150,00.asp
15. Many chips are automatically registered to a shelter or animal hospital. The shelter or hospital is contacted first, and it in turn contacts the owner. To bypass this procedure, owners need to register their information personally with the service provider (Homeagain or AVID, for example).
16. http://petplace.netscape.com/articles/artShow.asp?artID=960

17. Several other technologies have been developed for tagging objects with pointers. The HP Cooltown project, for example, tagged objects with infrared beacons that transmitted a URL directly to a PDA equipped with an infrared sensor and Web browser. Another, more recent development is the Semacode system, which enables camera phones to read specialized symbols (bar codes) that encode URLs. Using plain old light, the camera essentially takes a picture of the symbol and translates it into the URL with specialized software. The phone's Web browser then brings up the appropriate page with information about the tagged object.

18. http://www.rfidjournal.com/article/articleview/208

19. http://www.autoidcenter.org/

20. EPCglobal has created an interoperability protocol for RFID hardware called the Generation 2 standard. The standard would allow for interoperability among equipment suppliers and RF frequency standards around the world. There is some controversy over one equipment maker's (Intermec) demand for royalties, which could encourage other patent holders to follow suit and thereby stall innovation. Royalty fees would be passed on to customers, and implementations would slow down as costs rise. For now, most of the industry remains optimistic that Intermec and other companies will not jeopardize their own future gains.

21. The EPC namespace authority could be the Auto-ID Center or another standards organization chosen by the Auto-ID Center.

22. http://www.rfidjournal.com/article/articleview/208

23. http://www.rfidjournal.com/article/articleview/867/1/1/

24. http://www.rfidbuzz.com/wiki/Glossary/ONS?v=13yo

25. More detailed information about the EPC network and its application in the supply chain is at http://www.verisign.com/static/002109.pdf and http://archive.epcglobalinc.org/publishedresearch/MIT-AUTOID-WH-002.pdf.

26. http://www.inventoryops.com/RFIDupdate.htm

27. http://www.forrester.com/ER/Research/Brief/0,1317,33298,00.html

28. http://www.forrester.com/ER/Research/Brief/0,1317,33298,00.html

29. http://www.rfidjournal.com/article/articleview/867/1/1/

30. http://www.rfidjournal.com/article/articleview/867/1/1/

31. http://www.forrester.com/ER/Research/Brief/0,1317,33298,00.html

32. http://www.rfidjournal.com/article/articleprint/609/-1/1/

10

The X Internet Unleashes Real World Awareness Services Revolution

Navi Radjou is a Vice President in the Enter-prise Applications team at Forrester Research. Mr. Radjou's coverage areas include supply chain inte-gration and product life cycle management. Navi also investigates how globalized innovation is driv-ing new market structures and business processes. He was recently named by Supply&Demand Chain Executive magazine as one of the "Pros to Know," honoring an elite group of professionals who have excelled in improving the use of supply  *chain technology and practices within user companies. Navi received the award for introducing the term Adaptive Supply Networks (ASN), which describes "sense-and-respond" supply chains that are event driven, real-world aware, and self-regulating. Before joining Forrester, Navi was an IT Con-sultant in Asia and a Development Analyst at IBM's Toronto Software Lab. He earned his M.S. degree in information systems at Ecole Centrale Paris. He also attended the Yale School of Management.*

Navi Radjou
Vice President, Enterprise Applications, Forrester Research

XInternet technologies, like RFID and sensors, connect firms' IT systems to physical products, assets, and devices. Service-savvy corporations will exploit such Real World Awareness to transform their operations, processes, and business models—and to secure customer loyalty at the expense of rivals.

Most companies—from manufacturers to financial service firms—recognize the importance of customer service and understand the correlation between their internal operations and customer service. For instance, 81 percent of CIOs in service industries believe that customer acquisition and retention is one of their top priorities. And, 54 percent of supply chain executives view customer service improvement as one of their responsibilities. Hoping to boost their service performance, companies have invested in CRM applications and Internet portals. Some firms are bringing customer service to the next level, however, by extending the Internet to build linkages with potentially billions of physical assets, products, and devices (see Figure 10.1).

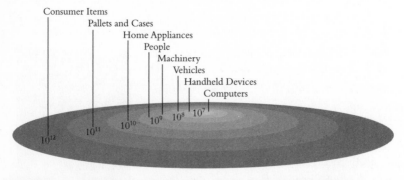

Physical object	Firms taking advantage of Internet connection with this object
Vehicles	**Norwich Union:** Its "pay as you drive" car insurance scheme uses a telematics system to charge drivers a monthly adjusted premium based on actual car usage.
Machinery	**Caterpillar:** Its MineStar system uses GPS to guide Cat machines in a mining field and uses data from embedded sensors to predict and prevent equipment failure. **Delta Air Lines:** Analyzes telemetry data from its aircraft engines to prevent failure.
Consumer items	**Michelin:** Uses RFID tags to track 20 million tires it has in transit in its supply chain.

Figure 10.1. Service-Savvy Leaders Are Using the Net to Connect to Billions of Physical Objects.

Some examples are shown in this list:

- **Michelin** is RFID-tagging its tires to deliver proactive customer service. Unlike its rivals, Michelin didn't wait for the enactment of the TREAD Act—a law passed by the U.S. Congress that requires auto suppliers to proactively deal with quality issues in tires—to start worrying about tire safety. In 2002, three years before the 2005 TREAD deadline, Michelin rolled out eTires. The eTires add-on sensor system measures the air pressure and temperature of commercial tires, which enables truck fleet operators to maximize asset use through reduced downtime and better fuel economy. Michelin is also using RFID tags to track, at any time, the 20 million tires it has in transit in its supply chain. Such extended visibility speeds the recall process.

- **Caterpillar** taps Global Positioning System (GPS) to make customers and dealers profitable. In 2000, Caterpillar rolled out Mine-Star, a GPS-enabled system that tracks in near real-time the location and status of all Cat machines in a mining field. By gaining insight into machine performance, mining firms prevent costly equipment failure and boost productivity while solidifying their loyalty to Cat. Cat dealers also benefit from Mine-Star because such value-added services carry profit margins of as much as 50 percent for heavy-equipment dealers versus their 10 percent margins from product sales.

- **Delta Air Lines** uses sensors to halve its maintenance costs. The top 10 U.S. airlines, including Delta, lose $858 million per year because of cancellations and delays, which are caused mainly by mechanical snafus, like engine failures. To prevent these losses, Delta uses SmartSignal's predictive analytics to spot early warning signs in daily engine performance data collected from its 550 planes. The upside? Delta can potentially save 51 percent, or $355 million, on its annual maintenance costs.

- **Norwich Union** relies on telematics to grow market share. Norwich is not resting on its laurels as the number-one provider in U.K. car insurance. To lure uninsured, low-risk drivers, the company is piloting the Pay As You Drive (PAYD) scheme, where the premium is based on how often, when, and where a customer uses her car. Three thousand volunteers are driving cars equipped with a black box linked to a telematics system, designed and operated by Orange UK and IBM, that tracks car usage patterns. Encouraged by PAYD's early results, Norwich plans to roll it out across the United Kingdom in 2004 and across Europe in 2005.

MOST FIRMS ARE DISCONNECTED FROM THE PHYSICAL WORLD—BUT NOT FOR LONG

Initiatives like Cat's MineStar and Norwich's PAYD led Forrester to find out whether mainstream firms collect enough data on the identity, location, and status of their physical assets and products. In Forrester's Business Technographics September 2003 North American Study, Forrester surveyed 172 North American executives, 43 percent of whom work in manufacturing firms and 57 percent in service and distribution firms (see Figure 10.2). It found that, across industries, most firms don't bother collecting in-depth data about their assets. Only 35 percent of firms told Forrester that they know the identity of their physical assets. Seventy-seven percent of firms don't track their asset locations, and 81 percent lack visibility into asset status. Forty-eight percent of firms don't see the business case for collecting in-depth asset data, yet only 22 percent feel that they will incur no penalty from not tracking their assets accurately. Interviewees believe that better asset visibility can help cut supply chain costs and improve product quality.

Corporations have only limited visibility into their products. Fifty-six percent of those surveyed say that they have in-depth identity data on their products. Forty-one percent have such data on their products' locations. Only 30 percent, however, can monitor the status of their products at any given time. Forty-four percent of firms don't see the need for collecting in-depth product data, and 20 percent feel that doing so would be too expensive. Half the firms surveyed recognize that a lack of in-depth product data can increase costs and slow response to market shifts—and 65 percent believe that better product tracking can improve customer service.

Firms with limited visibility into their assets and products have faced no negative consequences, so far. This situation is bound to change, however, as firms face growing pressure from government (which is introducing regulations like TREAD and C-TPAT), customers (Wal-Mart, for example), and competition to connect with the physical world. For instance, GE Aircraft Engines is stealing lucrative aftermarket service contracts from rivals like Rolls-Royce by closely monitoring how customers are using products from GE and its competitors. Rolls-Royce can't afford to lose these margin-rich product maintenance deals to GE.

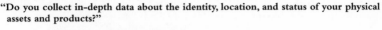

"Do you collect in-depth data about the identity, location, and status of your physical assets and products?"

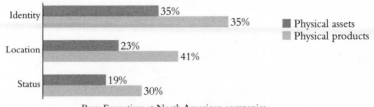

Base: Executives at North American companies

"Why don't you collect in-depth data about your physical assets and products?"

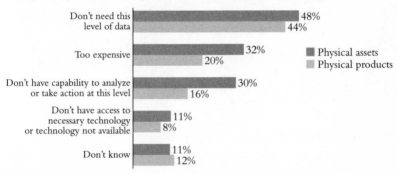

Base: Executives at companies that don't collect in-depth data about their assets or products (multiple responses accepted)

Figure 10.2. Few Firms Collect In-Depth Data about Physical Assets and Products.

X INTERNET LETS FIRMS REAP SERVICE PROFITS FROM TIES TO PHYSICAL WORLD

Growing pressure from regulators, customers, and competitors to build linkages with the physical world will drive firms across industries to embrace a new set of technologies that Forrester calls the Extended Internet, or X Internet (see Figure 10.3). Forrester defines *X Internet* as a set of technologies that connect firms' information systems to physical assets, products, and devices.

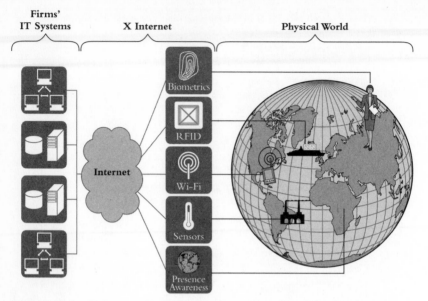

Figure 10.3. X Internet Technologies Boost Firms' Real World Awareness.

Smart firms will use the X Internet to boost their physical-world IQs and expand their revenue bases well beyond products to include more margin-rich services (see Figure 10.4). They will increase the Real World Awareness of their customer-facing processes by tapping technologies such as the ones in this list:

- **Biometrics,** to better identify employees, consumers, and partners. Growing identity theft and national security concerns have led firms to look beyond employee badges and Social Security numbers to uniquely identify their staff members and partners. Biometrics systems identify individuals by using biological attributes, such as fingerprints or retinal scans, that are hard to duplicate. For instance, Kroger stores use biometric payment systems.

- **RFID tags,** to pinpoint product location and content. U.S. manufacturers and retailers, which import $1.12 trillion goods per year, get only weekly notification on their shipment status. The result? Delays catch companies by surprise and force manufacturers to shut down just-in-time plants and retailers to scrap Christmas promotions. This situation is changing, however: Target, which imports $7.1 billion in goods, uses Savi Technology's RFID-based visibility application to track containers carrying its U.S.-bound shipments in near real-time.

- **Wi-Fi,** to unleash wireless access to enterprise applications. Firms want mobile workers to remain productive while they're away from their desks. One upshot is that 58 percent of North American firms are piloting or deploying Wi-Fi networks, which enable mobile access to enterprise software. Eastman Chemical, for example, enabled its 600-acre campus in Kingsport, Tennessee, with Wi-Fi so that its warehouse workers can track inventory on their PDAs while its engineers monitor chemical mixtures from their laptops.

- **Sensors,** to monitor asset usage and performance. Nippon Television cannot afford to have its lighting systems fail during a live show, such as "Sports MAX," and semiconductor leader TSMC can face $100,000 in lost revenues for every hour that its chip-making device is down. That's why equipment suppliers like Matsushita Electric Works and Applied Materials are embedding sensors into their wares to predict and prevent costly product failures well before they might occur.

- **Presence awareness,** to gain insight into the location of certain employees or partners. In emergency cases, firms need to quickly find out what's the best way to reach employees or partners. This process will be easier if individuals willingly share their electronic status (for example, "Am working from home, cell phone is off, IM me"). Presence-enabled communications promise such real-time interactions by contacting people via their choice of device or application. One example: When Ford Motor's paint booths are about to run out of paint, workers use GlobeStar Systems' ConnexALL to alert the first technician who is available on his cell phone.

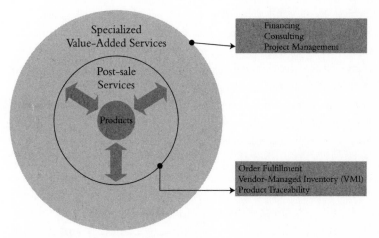

Figure 10.4. Smart Firms Will Exploit the X Internet to Capture Margin-Rich Service Revenues.

Firms' Service Culture and Goals Will Shape Their X Internet Initiatives

As the cost of X Internet technologies—such as sensors, drops, and standards like Electronic Product Code—matures, firms will need to decide how to effectively integrate them with their existing customer service processes and business models. To determine the scope of this organizational alignment, however, firms will first define one of these three goals of their X Internet initiatives:

* Functional enhancement
* Process optimization
* Business model reinvention

CULTURAL CONSERVATIVES WILL USE X INTERNET TO ACHIEVE FUNCTIONAL ENHANCEMENT

Irrespective of where a firm stands in its industry value chain—whether it's a supplier, an original equipment manufacturer, a dealer, or a customer—functional enhancement requires minimal investment and cultural change. In this scenario, firms will use the X Internet to squeeze efficiency from a specific function, whether it's R&D, sourcing, manufacturing, distribution, or sales and service (see Figure 10.5). In particular, capital-intensive firms will trim the total cost of asset ownership. A National Science Foundation study has shown that manufacturers can save 51 percent in asset maintenance by using X Internet-enabled predictive analytics software. Service firms will also use the X Internet to improve asset use. For instance, restaurant chains can't rapidly determine now whether the perishable items they have received are acceptable without first unloading all pallets within a shipment. That's not a problem, however, for freshness-obsessed seafood restaurants that use Sensitech's wireless sensors to log the sea-to-kitchen temperatures of their piscine assets.

As for commodity suppliers, they will wrap products with value-added services. Firms selling thin-margin products, such as bulk chemicals, will tap the X Internet to wrap their commodity goods with value-added services. The pharmaceutical distributor Cardinal Health, for example, didn't make $50 billion in 2002 from selling syringes but, rather, from selling X Internet-enabled services. As another example, Cardinal's Pyxis MedStations, a profile-based medical dispensing system, helps hospitals automate drug delivery to patients, by curbing errors, loss, and theft and saving nurses' time. Cardinal also replenishes the dispensers just in time, which saves hospitals inventory costs.

OEMs and dealers will use the X Internet to curb aftermarket service costs. OEMs and dealers that sell products with long life cycles, such as aircraft and cars, can make as much as 80 percent profit margins on aftermarket services—as long as they run lean operations, like keeping spares inventory low. Using Reynolds and Reynolds' telematics system, auto dealers, which make 50 percent of their profits on aftermarket products, can dynamically adjust their spares inventory levels by tracking customers' car usage in near real-time.

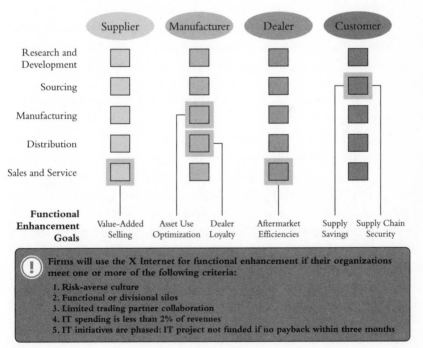

Figure 10.5. Conservative Firms Will Use X Internet for Functional Enhancement.

Finally, global buyers will boost supply chain security. Only 2 percent of the 21,000 containers arriving daily in U.S. ports are physically inspected. As one Fortune 50 executive noted, "If an act of terrorism were committed on one of our containers, it would be a company-ending event." As large U.S. importers like Wal-Mart and Dell seek to curb liability exposure, they will emulate Sara Lee and HP, which have joined RFID-enabled supply chain security initiatives, like Operation Safe Commerce and Smart and Secure Tradelanes. These initiatives can track in near real-time the location and status of cargo containers all the way from their origin to their arrival in U.S. ports.

Case in point: Cargill. Cargill Crop Nutrition rolled out the X Internet system Insite Variable Rate Nitrogen (VRN), a precision farming offering that uses GPS, sensors, and a GIS (geographic information system) tool from Linnet to create detailed digital maps of a farming field. Insite VRN helps Cargill dealers analyze these maps and recommend to farmers the right amount of nutrients to apply in the right place, which boosts crop yield and curbs overfertilization. By selling Insite VRN as a value-added service, Cargill dealers build customer intimacy while making 35 percent gross margins. Cargill's gain? Dealer loyalty.

FIRMS SEEKING FLEXIBILITY WILL USE X INTERNET TO OPTIMIZE END-TO-END PROCESSES

Caterpillar and General Electric lead the pack of companies that think in terms of end-to-end processes, not just functional silos. Such process-savvy firms with a risk-tolerant corporate DNA will use the X Internet to optimize cross-functional process flows and extend them to trading partners (see Figure 10.6). In particular, regulated manufacturers will nip product quality issues in the bud. Medical device makers learn too late about quality issues in their CT scanners or MRI machines, and when they do, they struggle to implement a corrective action—as required by the FDA—because they lack insight into the root cause. By feeding telemetry data collected from their medical devices into Agile Software's Product Service & Improvement application, however, firms like Hitachi Medical Corporation will not only be able to predict and prevent quality issues but also proactively comply with the FDA by redesigning or remanufacturing their products and avoid costly recalls.

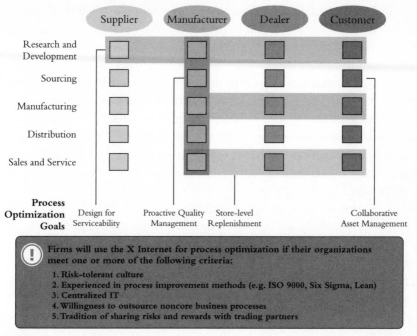

Figure 10.6. Efficiency-Seeking Firms Will Use X Internet to Build Service-Centric Business Processes.

CPG suppliers will proceed with store-level replenishment. Wal-Mart envisions its suppliers replenishing its individual outlets just in time, based on RFID-enabled store-level inventory data. To rapidly act on such store-level data, CPG suppliers, like Unilever, will roll out composite processes that synchronize the supply-side response to RFID-enabled demand signals by integrating activities like sales, logistics, and production. Similarly, resource-constrained users can use the X Internet to outsource asset management to third-party service providers. To cope with an aging workforce and the high costs of equipment downtime, industrial asset users will emulate IT asset users by outsourcing maintenance to OEMs and suppliers. Texas Instruments (TI), for instance, outsourced to Air Liquide the full management of industrial gas distribution systems used in TI's fabrication facilities. Air Liquide uses its FabView SCADA, industrial software designed to manage shop-floor processes, to remotely monitor the status and usage of TI's fabrication assets, which enables the use of predictive maintenance and just-in-time gas replenishment.

Case in point: The Department of Defense (DoD). To maximize defense preparedness and boost operational agility, the Department of Defense is rolling out performance-based logistics (PBL). Under PBL, the support of military assets—from weapon systems to aircraft carriers—is outsourced to suppliers for a guaranteed level of performance at the same or a reduced cost. PBL contractors commit to analyze sensor data from military assets to predict maintenance requirements and proactively line up resources to meet these needs. DoD projects that, by outsourcing support for its M1 Abrams tanks alone, it will save taxpayers $17 billion in maintenance over the next 30 years of the M1's service.

BOLD FIRMS WILL USE THE X INTERNET CRAFT A SERVICE-CENTRIC BUSINESS MODEL

Some companies like to go for bigger "bang," or maybe they are simply in a desperate situation. These companies will use the X Internet to reinvent their business models (see Figure 10.7).

In particular, expert asset users will become service providers. Half of ChevronTexaco's plant operators will be retiring over the next seven years, which will force the oil giant to outsource the maintenance of its refineries to third-party service providers. Eyeing this growing demand for outsourced support services, capital-intensive firms, like Duke Energy, that have mastered the art of using the X Internet for asset maintenance are spinning off their internal maintenance staff into an outsourced service firm.

Expect big-ticket product and service providers to let buyers pay as they go. Recession-hit manufacturers and airlines aren't inclined to cough up millions of dollars for stamping machines or aircraft engines. Rather than sell these big-ticket items as capital assets, suppliers will use the X Internet's ability to track product usage to charge for their products on a pay-per-use or pay-per-performance basis, as Rolls-Royce does with its Power-by-the-Hour program. Firms offering high-priced services, like insurance, will follow suit.

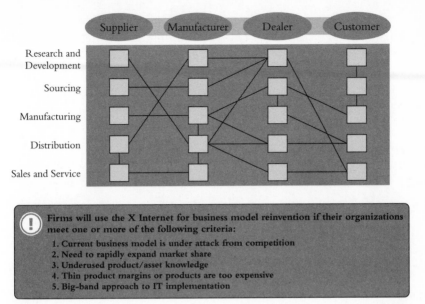

Figure 10.7. Aggressive Firms Will Use X Internet for Service-Centric Business Model Reinvention.

Finally, demand-driven firms will build adaptive supply networks. Freshly made cement concrete lasts only two hours, and apparel items are out of fashion before they hit the stores. Suppliers of these perishable or seasonal items need adaptive supply networks to cost-effectively address fickle demand. To keep up its 20-minute delivery window—nine times less than the industry average—cement supplier CEMEX tracks all its trucks' locations and the status of their shipments in near real-time. If a truck is heading toward a traffic jam or its concrete is becoming hard, for example, CEMEX either reroutes the truck or redirects it to a closer construction site.

Case in point: Delta Air Lines. As noted earlier in this chapter, Delta uses an X Internet-enabled predictive maintenance tool to reduce routine inspections and major repairs from 85 percent to 24 percent of total maintenance expenses. By cutting maintenance requirements on its own aircraft, Delta frees its 11,000 support staff members to service rival airlines. This practice, known as *in-sourcing,* brought Delta $150 million in revenues in 2002 and yielded 30 percent profit margins. Delta plans to grow its in-sourcing revenues to $500 million by 2005.

Index